FOUNDATIONS OF QUANTUM DYNAMICS

To my sons Michael and Stephen

FOUNDATIONS OF QUANTUM DYNAMICS

S. M. BLINDER
Department of Chemistry,
University of Michigan,
Ann Arbor, Michigan, U.S.A.

1974
ACADEMIC PRESS
LONDON NEW YORK SAN FRANCISCO
A Subsidiary of Harcourt Brace Jovanovich, Publishers

ACADEMIC PRESS INC. (LONDON) LTD.
24/28 Oval Road
London NW1

United States Edition published by
ACADEMIC PRESS. INC.
111 Fifth Avenue
New York, New York 10003

Library of Congress Catalog Card Number: 74-17426
ISBN:0-12-106050-0

PRINTED IN GREAT BRITAIN BY
PAGE BROS. (NORWICH) LTD., NORWICH

PREFACE

The subject of this monograph is the mathematical formalism of nonrelativistic quantum dynamics. The structure and foundations of this formalism are examined in detail with emphasis on fundamental principles rather than concrete applications. Still, our approach is slanted towards application to the domains of molecular physics and quantum chemistry. Our scope is thus limited to nonrelativistic wave mechanics and semiclassical radiation theory. We stop short of more advanced developments including the Dirac equation, quantum electrodynamics, formal scattering theory, density matrices and diagram techniques. These latter topics are, in fact, quite adequately covered in a number of current textbooks. I might suggest, however, that this monograph will prove of some value in bridging the gap between elementary wave mechanics and these more advanced techniques.

We begin, in Chapter 1, with a brief survey of some relevant topics in classical mechanics: Hamiltonian dynamics, Poisson brackets and Hamilton–Jacobi theory. Chapter 2 covers classical electrodynamics on the level of the Maxwell–Lorentz equations, emphasizing those aspects which pertain to radiative transitions in atoms and molecules. Special relativity is also developed as an outgrowth of electrodynamics. In subsequent chapters, points of connection between quantum-dynamical formalism and these classical theories are analyzed in detail. Chapter 3 reviews the postulates and general principles of quantum mechanics, including a few fine points which are usually glossed over in both elementary and advanced treatments. Chapter 4 is about time-dependent quantum mechanics, mainly the time-dependent Schrödinger equation. In chapter 5, the formalism of quantum dynamics is applied to the free particle. Some discussion is also given on continuum eigenstates and on the basic ideas of potential scattering. Chapter 6 is about Green's functions, both as mathematical tools used elsewhere and, in their own right, as the basis of high-powered computational formalism. Chapter 7 concerns transitions among quantum states and their treatment by time-dependent perturbation theory. In the final chapter we consider first the semiclassical theory of electromagnetic interactions. Our line of development then culminates in a fairly extensive discussion of radiative transitions in atomic and molecular systems. Two appendices give some required material on the Dirac deltafunction and Fourier analysis.

I should like to acknowledge the assistance of Dr. Eugene Lopata in critically reading parts of the manuscript. The comments of Professor Roy McWeeny and Professor David Craig were also extremely helpful.

S.M.B.
Ann Arbor, Michigan
July 1974

CONTENTS

1

Classical Dynamics*

In this chapter we shall briefly survey those parts of classical dynamics which have particular relevance to quantum theory. Principal among these are Hamiltonian formalism, equations of motion in Poisson bracket form and Hamilton–Jacobi theory.

1.1 Lagrange's Equations

It is assumed that the reader is familiar with the derivation of Lagrange's equations starting with Newton's equations of motion. We shall outline here an alternative derivation based on Hamilton's principle.

Let the dynamical structure of a classical system be characterized by some function $L(q, \dot{q}, t)$ of the n generalized coordinates q_i, n generalized velocities $\dot{q}_i$ $(i = 1 \dots n)$ and time t. The kinematical behaviour of the system is described by a *trajectory*, i.e. a set of functional relations

$$q_i = q_i(t), \qquad i = 1 \dots n \tag{1.1.1}$$

consistent with the constraints imposed on the system. The task is now to relate kinematics with dynamics. Assume that one can parametrically represent every conceivable trajectory connecting configuration $q_1^{(0)} \dots q_n^{(0)}$ at an initial time t_0 with configuration $q_1^{(1)} \dots q_n^{(1)}$ at a later time t_1. One of these paths will represent the *actual* motion of the system between times t_0 and t_1. *Hamilton's principle* is the postulate that the time integral of $L(q, \dot{q}, t)$ assumes a stationary value along the actual trajectory, i.e.,

$$\delta \int_{t_0}^{t_1} L(q, \dot{q}, t)\,\mathrm{d}t = 0, \tag{1.1.2}$$

in which δ represents variation of the parameters determining a trajectory. In terms of differential variations in the q_i and $\dot{q}_i$ at each point in time, one can also write

$$\int_{t_0}^{t_1} \sum_{i=1}^{n} \left(\frac{\partial L}{\partial q_i} \delta q_i + \frac{\partial L}{\partial \dot{q}_i} \delta \dot{q}_i \right) \mathrm{d}t = 0. \tag{1.1.3}$$

* For a more complete treatment of the subject, there is no finer reference than H. Goldstein, *Classical Mechanics* (Addison-Wesley, Cambridge, Massachusetts, 1951).

Noting that

$$\delta \dot{q}_i = \frac{\mathrm{d}}{\mathrm{d}t} \delta q_i \tag{1.1.4}$$

the second summation can be integrated by parts:

$$\int_{t_0}^{t_1} \frac{\partial L}{\partial \dot{q}_i} \frac{\mathrm{d}}{\mathrm{d}t} (\delta q_i) \, \mathrm{d}t = \frac{\partial L}{\partial \dot{q}_i} \delta q_i \Big]_{t_0}^{t_1} - \int_{t_0}^{t_1} \frac{\mathrm{d}}{\mathrm{d}t} \left(\frac{\partial L}{\partial \dot{q}_i} \right) \delta q_i \, \mathrm{d}t. \tag{1.1.5}$$

The boundary terms vanish since, by supposition, the initial and final configurations of the system are specified, i.e.

$$\delta q_i^{(0)} = 0, \quad \delta q_i^{(1)} = 0. \tag{1.1.6}$$

Hamilton's principle thereby reduces to

$$\int_{t_0}^{t_1} \sum_{i=1}^{n} \left\{ \frac{\partial L}{\partial q_i} - \frac{\mathrm{d}}{\mathrm{d}t} \left(\frac{\partial L}{\partial \dot{q}_i} \right) \right\} \delta q_i \, \mathrm{d}t = 0. \tag{1.1.7}$$

Now if the δq_i can be arbitrarily varied *independently of one another* (holonomic system), each curly bracket must individually vanish. We arrive thus at *Lagrange's equations of motion*

$$\frac{\mathrm{d}}{\mathrm{d}t} \frac{\partial L}{\partial \dot{q}_i} - \frac{\partial L}{\partial q_i} = 0, \quad i = 1 \dots n. \tag{1.1.8}$$

Solution of these equations, consistent with a set of specified initial conditions, say, $q_i^{(0)}$ and $\dot{q}_i^{(0)}$ $(i = 1 \dots n)$, suffices in principle to determine each generalized coordinate as a function of time (eqn 1.1.1).

We have yet however to specify the functional form of the *Lagrangian* $L(q, \dot{q}, t)$. It is suggestive to consider the motion of a particle in a conservative field of force. By Newton's second law,

$$m\ddot{\mathbf{r}} = -\nabla V(\mathbf{r}), \tag{1.1.9}$$

noting that conservative forces can be represented as the negative gradient of a potential energy. In cartesian coordinates, the equations of motion take the form

$$m\ddot{x}_i + \frac{\partial V}{\partial x_i} = 0, \qquad i = 1, 2, 3. \tag{1.1.10}$$

It is easily seen that these are isomorphous with (1.1.8) provided that one identifies

$$L(\mathbf{r}, \dot{\mathbf{r}}) = \tfrac{1}{2} m\dot{\mathbf{r}}^2 - V(\mathbf{r}). \tag{1.1.11}$$

In the great majority of cases, the Lagrangian is simply the difference between the kinetic and potential energies

$$L = T - V. \tag{1.1.12}$$

Should there arise some ambiguity in defining T or V, however, one has recourse to the fundamental significance of the Lagrangian as that function which establishes (1.1.8) as the equations of motion governing the system.

Consider, for example, the equation for a damped harmonic oscillator:

$$m\ddot{x} + \eta\dot{x} + kx = 0. \tag{1.1.13}$$

Because of the velocity-dependent dissipative force $-\eta\dot{x}$, the Lagrangian *cannot* be represented in the form (1.1.12). It is easily verified however that (1.1.13) can be derived from a time-dependent Lagrangian†

$$L(x, \dot{x}, t) = e^{\gamma t}(\tfrac{1}{2}m\dot{x}^2 - \tfrac{1}{2}kx^2), \qquad \gamma \equiv \eta/m. \tag{1.1.14}$$

1.2 Hamiltonian Dynamics

For conservative systems, in which the Lagrangian contains no explicit time dependence, i.e.

$$\partial L/\partial t = 0, \tag{1.2.1}$$

an important conservation principle applies. To demonstrate this, multiply each Lagrange equation (1.1.8) by the corresponding $\dot{q}_i$ and sum:

$$\sum_i \left(\dot{q}_i \frac{d}{dt}\frac{\partial L}{\partial \dot{q}_i} - \dot{q}_i \frac{\partial L}{\partial q_i} \right) = 0. \tag{1.2.2}$$

Note now that

$$\frac{dL}{dt} = \sum_i \left(\frac{\partial L}{\partial q_i}\dot{q}_i + \frac{\partial L}{\partial \dot{q}_i}\ddot{q}_i \right) \tag{1.2.3}$$

and

$$\frac{d}{dt}\sum_i \frac{\partial L}{\partial \dot{q}_i}\dot{q}_i = \sum_i \left(\dot{q}_i \frac{d}{dt}\frac{\partial L}{\partial \dot{q}_i} + \frac{\partial L}{\partial \dot{q}_i}\ddot{q}_i \right). \tag{1.2.4}$$

Subtracting (1.2.3) from (1.2.4) and substituting into (1.2.2), we find

$$\frac{d}{dt}\left(\sum_i \frac{\partial L}{\partial \dot{q}_i}\dot{q}_i - L \right) = 0. \tag{1.2.5}$$

† For systems subject to forces which are not entirely conservative, Lagrange's equations can be generalized to

$$\frac{d}{dt}\frac{\partial L}{\partial \dot{q}_i} - \frac{\partial L}{\partial q_i} = Q_i, \qquad i = 1 \dots n \tag{1.1.8'}$$

which Q_i is the dissipative force associated with the ith degree of freedom. Thus, (1.1.13) can alternatively be represented using

$$L = \tfrac{1}{2}m\dot{x}^2 - \tfrac{1}{2}kx^2, \qquad Q = -\eta\dot{x}. \tag{1.1.14'}$$

Thus the *Hamiltonian function*

$$H \equiv \sum_i \frac{\partial L}{\partial \dot{q}_i} \dot{q}_i - L \tag{1.2.6}$$

is constant in time for a conservative system.

When L contains explicit time dependence, then H will as well. Equation (1.2.6) represents, in fact, a *Legendre transformation* in which the $\dot{q}_i$ are displaced as fundamental variables by the quantities $\partial L/\partial \dot{q}_i$. This is made explicit by taking the total differential of (1.2.6), noting that L is a function of the q_i, $\dot{q}_i$ and t:

$$\mathrm{d}H = \sum_i \left\{ \dot{q}_i \, \mathrm{d}\left(\frac{\partial L}{\partial \dot{q}_i}\right) + \frac{\partial L}{\partial \dot{q}_i} \mathrm{d}\dot{q}_i \right\} - \sum_l \left\{ \frac{\partial L}{\partial q_i} \mathrm{d}q_i + \frac{\partial L}{\partial \dot{q}_i} \mathrm{d}\dot{q}_i \right\} - \frac{\partial L}{\partial t} \mathrm{d}t. \tag{1.2.7}$$

The terms in $\mathrm{d}\dot{q}_i$ cancel, thus indicating that the new function H depends on the variables q_i, $\partial E/\partial \dot{q}_i$ and t.

The new variables

$$p_i \equiv \partial L/\partial \dot{q}_i \tag{1.2.8}$$

are known as *generalized momenta.* In cartesian coordinates, (1.2.8) reduces to the usual definition of linear momentum. For example, with (1.1.11),

$$p_x = \frac{\partial L}{\partial \dot{x}} = m\dot{x} \tag{1.2.9}$$

The corresponding pairs of generalized coordinates and momenta q_i, p_i are known as *conjugate variables.* Their product invariably has dimension ml^2/t, known as *action.* Thus, length and linear momentum, angle and angular momentum, time and energy are the common pairs of conjugate variables. In quantum theory, the fundamental constant $\hbar$ has dimensions of action while each pair of conjugate variables is governed by an uncertainty relation (cf Section 3.6A).

One can now write

$$H(q, p, t) \equiv \sum_i p_i \dot{q}_i - L(q, \dot{q}, t) \tag{1.2.10}$$

and, using (1.2.8) in (1.2.7),

$$\mathrm{d}H = \sum_i (\dot{q}_i \, \mathrm{d}p_i - \dot{p}_i \, \mathrm{d}q_i) - \frac{\partial L}{\partial t} \mathrm{d}t. \tag{1.2.11}$$

Evidently,

$$\partial H/\partial p_i = \dot{q}_i, \quad \partial H/\partial q_i = -\dot{p}_i, \qquad i = 1 \ldots n \tag{1.2.12}$$

and

$$\partial H/\partial t = -\partial L/\partial t. \tag{1.2.13}$$

Equations (1.2.12) are *Hamilton's equations of motion.* Usually the second of these is equivalent Newton's second law in the form

$$\dot{p}_i = F_i \tag{1.2.14}$$

while the first relates p_i to the generalized velocity. Equations (1.2.12) are also known, because of their theoretical importance, as *canonical equations of motion.* The q_i, p_i entering into these equations are correspondingly denoted *canonical variables.*

Comparison of (1.1.8) with (1.2.13) shows that a mechanical problem can be formulated either as a set of n second-order differential equations (Lagrange's) or as a set of $2n$ first-order differential equations (Hamilton's). In either case, $2n$ initial or boundary conditions are required for a complete solution.

Except under some rather exotic circumstances, the Hamiltonian function represents the total energy of a system†. Consider a Lagrangian in the form

$$L = T - V_0 - V_1 \tag{1.2.15}$$

in which the kinetic energy is a homogeneous quadratic function of the generalized velocities which does *not* involve the time explicitly, i.e.

$$T = \tfrac{1}{2}\sum_{i,j} a_{ij}(q)\dot{q}_i\dot{q}_j \tag{1.2.16}$$

while the potential energy, which may depend on t, contains (at most) a homogeneous linear dependence on generalized velocities, i.e.

$$V_0 = V_0(q, t), \qquad V_1 = \sum_i b_i(q, t)\dot{q}_i. \tag{1.2.17}$$

By Euler's theorem on homogeneous functions, we have

$$\sum_i \frac{\partial T}{\partial \dot{q}_i}\dot{q}_i = 2T \quad \text{and} \quad \sum_i \frac{\partial V}{\partial \dot{q}_i}\dot{q}_i = V. \tag{1.2.18}$$

† For the damped harmonic motion represented by the Lagrangian (1.1.14), the general prescription for constructing the Hamiltonian gives

$$p = \frac{\partial L}{\partial \dot{x}} = e^{\gamma t} m\dot{x} \tag{1.2.21}$$

and thus

$$H(x, p, t) = e^{-\gamma t}\frac{p^2}{2m} + e^{\gamma t}\frac{kx^2}{2}. \tag{1.2.22}$$

The sum of the kinetic and potential energies is however

$$E = \tfrac{1}{2}m\dot{x}^2 + \tfrac{1}{2}kx^2 \tag{1.2.23}$$

so that $H = e^{\gamma t}E$—one of those pathological instances in which $H \neq E$. If, however, one defines the Lagrangian in accordance with (1.1.8′) and (1.1.14′), a Hamiltonian equal to (1.2.23) is obtained.

Putting (1.2.15) into (1.2.6) and making use of (1.2.18), we find that

$$H = T + V_0 \equiv E \tag{1.2.19}$$

(V_1 having cancelled out). Under the very general circumstances considered, the Hamiltonian represents the sum of the ordinary kinetic and potential energies, hence the total energy of the system. When V_0 is time dependent, then E, as defined, will also be time dependent. When V_0 is independent of time, then, in accordance with (1.2.5), the energy of the system is conserved.

For a particle in a conservative field, application of (1.2.9) and (1.2.10) to the Lagrangian (1.1.11) leads to the Hamiltonian function

$$H(\mathbf{r}, \mathbf{p}) = \frac{p^2}{2m} + V(\mathbf{r}) \tag{1.2.20}$$

This is representative of a large number of instances in which the Hamiltonian can be constructed simply by expressing the energy as a function of generalized coordinates and momenta.

The Hamiltonian formulation of mechanics outlined in this section is entirely equivalent in physical content to the Lagrangian formulation. In fact, Hamilton's equations (1.2.13) often reduce to the very same differential equations as Lagrange's equations (1.1.8). Hamiltonian dynamics nonetheless possess a number of conceptual advantages. One is the physical significance of the Hamiltonian function itself. A second attractive feature is the near symmetry in the roles of generalized coordinates and momenta. These can accordingly be employed with greater flexibility than the variables in Lagrangian formalism. In consequence, Hamiltonian dynamics plays an important role in the formulation of both quantum theory and statistical mechanics.

1.3 Equations of Motion for Dynamical Variables

Let $A(q, p, t)$ represent some physical property of the dynamical system expressed as a function of canonical variables and time. As the system evolves in conformity with Hamilton's equations, each dynamical variable A will correspondingly vary with time in accordance with

$$\frac{\mathrm{d}A}{\mathrm{d}t} = \frac{\partial A}{\partial t} + \sum_{i=1}^{n} \left(\frac{\partial A}{\partial q_i} \dot{q}_i + \frac{\partial A}{\partial p_i} \dot{p}_i \right). \tag{1.3.1}$$

By substituting for $\dot{q}_i$ and $\dot{p}_i$ from (1.2.13) we obtain

$$\frac{\mathrm{d}A}{\mathrm{d}t} = \frac{\partial A}{\partial t} + \sum_{i=1}^{n} \left(\frac{\partial A}{\partial q_i} \frac{\partial H}{\partial p_i} - \frac{\partial A}{\partial p_i} \frac{\partial H}{\partial q_i} \right). \tag{1.3.2}$$

It is useful to define the *Poisson bracket* of two dynamical variables A and

B as follows:

$$\{A, B\} \equiv \sum_{i=1}^{n} \left(\frac{\partial A}{\partial q_i} \frac{\partial B}{\partial p_i} - \frac{\partial B}{\partial q_i} \frac{\partial A}{\partial p_i} \right). \tag{1.3.3}$$

It can be shown that this quantity is invariant to alternative choices of the canonical variables in a given system. Since Poisson brackets provide an important link to quantum mechanics, we shall summarize some of their properties. The following identities are readily demonstrated:

$$\{A, A\} = 0 \tag{1.3.4}$$

$$\{A, B\} = -\{B, A\} \tag{1.3.5}$$

$$\{A, B + C\} = \{A, B\} + \{A, C\} \tag{1.3.6}$$

$$\{A^2, B\} = 2A\{A, B\} \tag{1.3.7}$$

$$\{A, BC\} = C\{A, B\} + B\{A, C\} \tag{1.3.8}$$

and

$$\{\{A, B\}, C\} + \{\{B, C\}, A\} + \{\{C, A\}, B\} = 0. \tag{1.3.9}$$

(Jacobi's identity)

When A and B are canonical variables themselves, we find

$$\{q_i, q_j\} = 0, \quad \{p_i, p_j\} = 0 \qquad \text{all } i, j \tag{1.3.10}$$

but

$$\{q_i, p_j\} = \delta_{ij}. \tag{1.3.11}$$

Returning now to the equation of motion (1.3.2), we recognize that it can be written

$$\frac{\mathrm{d}A}{\mathrm{d}t} = \frac{\partial A}{\partial t} + \{A, H\}. \tag{1.3.12}$$

This points up that, besides its significance as the energy, the Hamiltonian also governs the time development of all dynamical variables. When applied to the canonical variables, (1.3.12) reproduces Hamilton's equations in Poisson bracket form:

$$\dot{q}_i = \{q_i, H\} \tag{1.3.13}$$

and

$$\dot{p}_i = \{p_i, H\}. \tag{1.3.14}$$

When A is the Hamiltonian itself, then, because of (1.3.4),

$$\frac{\mathrm{d}H}{\mathrm{d}t} = \frac{\partial H}{\partial t}. \tag{1.3.15}$$

Thus $H(q, p)$ with no *explicit* dependence on time will maintain a constant value even as the coordinates and momenta vary with time in the course of the system's motion. This is consistent, of course, with the conservation of energy in a conservative system.

A dynamical variable for which $dA/dt = 0$ is known as a *constant of the motion*. According to (1.3.12), a constant of the motion must fulfil two conditions:

$$\partial A/\partial t = 0 \quad \text{and} \quad \{A, H\} = 0. \tag{1.3.16}$$

1.4 Hamilton–Jacobi Theory

We shall present an *ad rem* derivation of the Hamilton–Jacobi equation which does not explicitly develop its connection with either canonical transformations or variational principles. *Hamilton's principal function* (sometimes called, rather loosely, the *action integral*) is defined as

$$S(q, q^{(0)}, t) \equiv \int_0^t L(q', \dot{q}', t')\, dt'. \tag{1.4.1}$$

It is presumed that the integrand describes an *actual* trajectory of the system during the time interval 0 to t, such that $\delta S = 0$, in accord with Hamilton's principle. The trajectory is characterized by the set of functional relations

$$q_i' = q_i'(t'), \quad \dot{q}_i' = \dot{q}_i'(t'), \quad i = 1 \ldots n, \tag{1.4.2}$$

obtained by solution of the equations of motion subject to the initial and final conditions

$$q_i'^{(0)} = q_i^{(0)}, \; q_i'(t) = q_i, \quad i = 1 \ldots n. \tag{1.4.3}$$

Incorporating (1.4.3) very explicitly into (1.4.2), we can write

$$q_i' = q_i'(q, q^{(0)}, t, t'), \quad \dot{q}_i' = \dot{q}_i'(q, q^{(0)}, t, t'), \quad i = 1 \ldots n \tag{1.4.4}$$

and correspondingly,

$$S(q, q^{(0)}, t) = \int_0^t L[q'(q, q^{(0)}, t, t'), \; \dot{q}'(q, q^{(0)}, t, t'), t']\, dt'. \tag{1.4.5}$$

Now

$$\frac{\partial S}{\partial q_i} = \int_0^t \sum_{j=1}^n \left(\frac{\partial L}{\partial q_j'} \frac{\partial q_j'}{\partial q_i} + \frac{\partial L}{\partial \dot{q}_j'} \frac{\partial \dot{q}_j'}{\partial q_i} \right) dt'. \tag{1.4.6}$$

Noting that

and

$$\left.\begin{aligned} \frac{\partial L}{\partial q'_j} &= p'_j \\ \frac{\partial L}{\partial \dot{q}'_j} &= \frac{\mathrm{d}}{\mathrm{d}t}\frac{\partial L}{\partial \dot{q}'_j} = \dot{p}'_j \end{aligned}\right\} \tag{1.4.7}$$

we can write

$$\frac{\partial S}{\partial q_i} = \int_0^t \frac{\mathrm{d}}{\mathrm{d}t}\left(\sum_j p'_j \frac{\partial q'_j}{\partial q_i}\right) \mathrm{d}t'$$

$$= \sum_j p'_j \frac{\partial q'_j}{\partial q_i}\Bigg]_0^t = \sum_j \left(p_j \frac{\partial q_j}{\partial q_i} - p_j^{(0)} \frac{\partial q_j^{(0)}}{\partial q_i}\right). \tag{1.4.8}$$

Thus

$$\partial S/\partial q_i = p_i, \quad i = 1 \dots n. \tag{1.4.9}$$

Analogously, it is shown that

$$\partial S/\partial q_i^{(0)} = -p_i^{(0)}, \quad i = 1 \dots n \tag{1.4.10}$$

Consider now the time derivative of S:

$$\frac{\partial S}{\partial t} = L(q, \dot{q}, t) + \int_0^t \sum_i \left(\frac{\partial L}{\partial q'_i}\frac{\partial q'_i}{\partial t} + \frac{\partial L}{\partial \dot{q}'_i}\frac{\partial \dot{q}'_i}{\partial t}\right) \mathrm{d}t'. \tag{1.4.11}$$

The last integral can analogously be transformed to

$$\sum_i p'_i \frac{\partial q'_i}{\partial t}\Bigg]_0^t \tag{1.4.12}$$

Applying the identity

$$\left(\frac{\partial q'_i}{\partial t}\right)_{q_i} = -\left(\frac{\partial q_i}{\partial t}\right)_{q'_i}\left(\frac{\partial q'_i}{\partial q_i}\right)_t \tag{1.4.13}$$

this becomes

$$-\sum_i \left(p_i \dot{q}_i - p_i^{(0)} \dot{q}_i^{(0)} \frac{\partial q_i^{(0)}}{\partial q_i}\right) \tag{1.4.14}$$

Thus

$$\frac{\partial S}{\partial t} = L - \sum_i p_i \dot{q}_i = -H. \tag{1.4.15}$$

Combining with (1.4.9), we obtain the time-dependent Hamilton–Jacobi equation

$$\frac{\partial S}{\partial t} + H\left(q, \frac{\partial S}{\partial q}, t\right) = 0. \tag{1.4.16}$$

Hamilton–Jacobi theory provides an alternative formulation of classical mechanics on a highly abstract and sophisticated level. Hamilton's principal function, in conjunction with the $2n$ equations (1.4.9) and (1.4.10), suffices, in principle, to determine the $q_i(t)$ and $p_i(t)$ and thus to solve the mechanical problem. For our purposes, however, we need only determine $S(q, q^{(0)}, t)$ from (1.4.5) by making use of known solutions of the equations of motion.

As a first illustration, consider the free particle in one dimension. The particle moves at a constant velocity v which can be expressed

$$v = \frac{x - x_0}{t}. \tag{1.4.17}$$

Thus the Lagrangian is

$$L = \frac{m}{2}\frac{(x - x_0)^2}{t^2} \tag{1.4.18}$$

and integration of (1.4.5) gives simply

$$S(x_1 x_0, t) = m(x - x_0)^2/2t. \tag{1.4.19}$$

Analogously for the free particle in three dimensions

$$S(\mathbf{r}, \mathbf{r}_0, t) = m(\mathbf{r} - \mathbf{r}_0)^2/2t. \tag{1.4.20}$$

As a second example, consider the linear harmonic oscillator. Write the Lagrangian

$$L(x, \dot{x}) = \frac{m}{2}(\dot{x}^2 - \omega^2 x^2) \tag{1.4.21}$$

in terms of the oscillation frequency

$$\omega^2 = k/m. \tag{1.4.22}$$

Specifically, as a function of t',

$$L(t') = \frac{m}{2}\left[\left(\frac{dx}{dt'}\right)^2 - \omega^2 x(t')^2\right] \tag{1.4.23}$$

in which

$$x(t') = A \sin \omega t' + B \cos \omega t' \tag{1.4.24}$$

from solution of the equation of motion. Consistent with the requisite initial and final conditions

$$x(0) = x_0 \qquad \text{and} \qquad x(t) = x \tag{1.4.25}$$

we have

$$A = \frac{x - x_0 \cos \omega t}{\sin \omega t}, \qquad B = x_0. \tag{1.4.26}$$

Putting (1.4.24) into (1.4.23) results in

$$L(t') = \frac{m\omega^2}{2}[(A^2 - B^2)\cos 2\omega t' - 2AB \sin 2\omega t']. \tag{1.4.27}$$

After integration and more trigonometric transformation

$$S(x, x_0, t) = \frac{m\omega}{2}[(A^2 - B^2)\sin \omega t \cos \omega t - 2AB \sin^2 \omega t]. \tag{1.4.28}$$

Finally, using (1.4.26) and rearranging, we obtain

$$S(x, x_0, t) = \frac{m\omega}{2 \sin \omega t}[(x^2 + x_0^2)\cos \omega t - 2xx_0]. \tag{1.4.29}$$

Note that this reduces to the free-particle result (1.4.19) in the limit $\omega \to 0$. The principal functions (1.4.19) and (1.4.29) will be required in Section 6.5 as adjuncts to the construction of Green's functions.

From the structure of $S(q, q^{(0)}, t)$ as an integral, the following identities are easily demonstrated:

$$S(q^{(0)}, q^{(0)}, 0) = 0 \tag{1.4.30}$$

and

$$S(q, q^{(0)}, t) = -S(q^{(0)}, q, -t) \tag{1.4.31}$$

$$S(q, q^{(0)}, t) - S(q', q^{(0)}, t') = S(q, q', t - t'). \tag{1.4.32}$$

For Hamiltonian independent of time, Hamilton's principal function is separable as follows:

$$S(q, q^{(0)}, t) = W(q, q^{(0)}) - Et \tag{1.4.33}$$

The function $W(q, q^{(0)})$, known as *Hamilton's characteristic function*, is then determined by the time-independent Hamilton–Jacobi equation

$$H\left(q, \frac{\partial W}{\partial q}\right) = E \tag{1.4.34}$$

in which the constant E is identified as the energy. Solution of (1.4.16) will, in general, yield n constants of the motion. These can often be chosen as the initial values of the generalized momenta: $p_1^{(0)} \ldots p_n^{(0)}$. The alternative representations $S(q, p^{(0)}, t)$ and $S(q, q^{(0)}, t)$ for Hamilton's principal function can be related by a Legendre transformation

$$S(q, q^{(0)}, t) = S(q, p^{(0)}, t) - \sum_{i=1}^{n} q_i^{(0)} p_i^{(0)} \tag{1.4.35}$$

dependence on the $p_i^{(0)}$ dropping out by virtue of (1.4.10). It follows moreover that

$$\frac{\partial S}{\partial p_i^{(0)}} = q_i^{(0)} \qquad i = 1 \ldots n. \tag{1.4.36}$$

To illustrate, the 3-dimensional free-particle Hamilton–Jacobi equation

$$\frac{\partial S}{\partial t} + \frac{1}{2m}\left[\left(\frac{\partial S}{\partial x}\right)^2 + \left(\frac{\partial S}{\partial y}\right)^2 + \left(\frac{\partial S}{\partial z}\right)^2\right] = 0 \tag{1.4.37}$$

easily yields the solution

$$S(\mathbf{r}, \mathbf{p}, t) = \mathbf{p} \cdot \mathbf{r} - Et \tag{1.4.38}$$

with

$$E = p^2/2m. \tag{1.4.39}$$

Now, by (1.4.35),

$$S(\mathbf{r}, \mathbf{r}_0, t) = S(\mathbf{r}, \mathbf{p}, t) - \mathbf{r} \cdot \mathbf{p}. \tag{1.4.40}$$

If dependence on **p** is now eliminated using (1.4.36) on (1.4.38)

$$v_i^{(0)} = \frac{\partial S}{\partial p_i} = x_i - p_i t/m \qquad i = 1, 2, 3, \tag{1.4.41}$$

the result is (1.4.20).

Hamilton–Jacobi theory suggests deep analogies between particle mechanics and optical theory†. Briefly, the basis of this is the fact that the momentum is a particle is represented by the gradient of S. Thus particle trajectories must run normal to surfaces of constant S, in analogy with the relationship between rays and wavefronts.

† A detailed discussion is given in Goldstein, *op cit*, p 307ff.

2

Electrodynamics*

2.1 Maxwell–Lorentz Equations

Maxwell in 1864 formalized all known electric and magnetic phenomena in terms of a set of four differential equations. Lorentz reformulated these as *microscopic* field equations, appropriate to the study of matter on the atomic scale. Lorentz's formulation assumes the absence of ponderable dielectric and magnetic media ($\varepsilon = 1, \mu = 1$) so that all charges and currents must be explicitly accounted for. The **D** and **E** fields are then equal, as are the **B** and **H** fields. We shall employ throughout the *gaussian* system of units† which is most convenient in atomic applications.

The four Maxwell–Lorentz equations determine and correlate the electric field $\mathbf{E}(\mathbf{r}, t)$ and the magnetic field $\mathbf{H}(\mathbf{r}, t)$ produced by a system of electrical charges and currents. These sources are specified by a charge density $\rho(\mathbf{r}, t)$ and a current density $\mathbf{j}(\mathbf{r}, t)$. In convective charge flux, these densities are related by

$$\mathbf{j} = \rho \mathbf{v} \tag{2.1.1}$$

in which $\mathbf{v}(\mathbf{r}, t)$ represents the instantaneous local velocity of charge.

The first Maxwell–Lorentz equation

$$\nabla \cdot \mathbf{E} = 4\pi\rho \tag{2.1.2}$$

represents a differential statement of Gauss' law: Every positive charge is a source of electric lines of force, every negative charge, a sink, the net number of lines of force emanating from an element of volume being equal to 4π times the net charge within the volume. Gauss' law derives, of course, from Coulomb's law, so that eqn (2.1.2) encompasses all electrostatic phenomena.

The second equation,

$$\nabla \cdot \mathbf{H} = 0 \tag{2.1.3}$$

* Excellent treatises on electromagnetic theory are J. D. Jackson, *Classical Electrodynamics* (John Wiley and Sons, New York, 1962) and Panofsky and Phillips, *Classical Electricity and Magnetism* (Addison-Wesley, Reading, Mass., 1962).

† Formulas in SI units can be converted to their guassian equivalents by the substitutions: $\varepsilon_0 \to \frac{1}{4}\pi$, $\mu_0 \to 4\pi/c^2$, $\mathbf{B} \to \mathbf{B}/c$.

is the magnetic analog of Gauss' law: Based on the apparent non-existence of free magnetic poles, the number of magnetic lines of force entering every element of volume equals the number emerging.

The third equation,

$$\nabla \times \mathbf{E} + \frac{1}{c}\frac{\partial \mathbf{H}}{\partial t} = 0 \tag{2.1.4}$$

is a differential statement of Faraday's law of electromagnetic induction, whereby a time-varying magnetic field gives rise to an electric field. This principle underlies the operation of electrical generators, electromotive force being induced by periodic motion of conductors through inhomogeneous magnetic fields.

Oersted first discovered that an electric current can produce a magnetic field. Quantitatively, this is governed by Ampère's law (also known as the Biot–Savart law), which can be stated in differential form

$$\nabla \times \mathbf{H} = \frac{4\pi}{c}\mathbf{j}. \tag{2.1.5}$$

Maxwell showed that Ampère's law is incomplete by the following argument. Taking the divergence of eqn (2.1.5),

$$\nabla \cdot \nabla \times \mathbf{H} = \frac{4\pi}{c}\nabla \cdot \mathbf{j}. \tag{2.1.6}$$

But, by a well-known vector identity, the left-hand side is zero. Ampère's law thereby implies that

$$\nabla \cdot \mathbf{j} = 0, \tag{2.1.7}$$

according to which no currents can originate from or terminate in any element of volume. This is indeed valid for steady-state phenomena. But, more generally, currents *can* be created and destroyed—at the plates of capacitors or in vacuum-tube space charge, for example. The requisite generalization of (2.1.7) is the *equation of continuity*

$$\partial\rho/\partial t + \nabla \cdot \mathbf{j} = 0. \tag{2.1.8}$$

This general relationship of hydrodynamic theory applies to any fluid whose overall quantity is conserved. According to (2.1.8), in every element of volume, the net influx or outflux of fluid is exactly balanced by a corresponding increase or decrease in fluid density. If the fluid has sources or sinks, the equation of continuity generalizes to

$$\partial\rho/\partial t + \nabla \cdot \mathbf{j} = S, \tag{2.1.9}$$

in which S represents the source density: the net quantity of fluid created per unit time per unit volume.

Applied to electrical charge, (2.1.8) means that a net current into any element of volume is necessarily correlated with an accumulation of charge therein.

It is accordingly suggested that (2.1.6) be generalized to

$$\nabla \cdot \nabla \times \mathbf{H} = \frac{4\pi}{c}\left(\nabla \cdot \mathbf{j} + \frac{\partial \rho}{\partial t}\right). \tag{2.1.10}$$

From (2.1.2) we have

$$\frac{\partial \rho}{\partial t} = \frac{1}{4\pi} \nabla \cdot \frac{\partial \mathbf{E}}{\partial t}, \tag{2.1.11}$$

so that

$$\nabla \cdot (\nabla \times \mathbf{H}) = \nabla \cdot \left(\frac{4\pi}{c}\mathbf{j} + \frac{1}{c}\frac{\partial \mathbf{E}}{\partial t}\right). \tag{2.1.12}$$

This result, as it stands, is trivial since it says nothing more than $0 = 0$. Although it is not rigorously implied that the operands in (2.1.12) are equal, Maxwell postulated that they were. Ampère's law (2.1.5), modified by addition of the *displacement current*,† constitutes the fourth Maxwell–Lorentz equation:

$$\nabla \times \mathbf{H} = \frac{4\pi}{c}\mathbf{j} + \frac{1}{c}\frac{\partial \mathbf{E}}{\partial t}. \tag{2.1.13}$$

The electromagnetic field equations, thus constituted, correctly predict the existence of electromagnetic radiation; this would not be so without Maxwell's conjecture on displacement current.‡

The first and fourth of the Maxwell–Lorentz equations—(2.1.2) and (2.1.13), respectively—are *source equations*: these determine the electric and magnetic fields produced by a distribution of charges and currents. The second and third equations—(2.1.3) and (2.1.4)—are homogeneous field equations: these determine the general mathematical structure of the fields. Upon further analysis, **E** and **H** can be expressed in terms of one vector and one scalar field.

2.2 Electromagnetic Potentials

This reduction is based upon two vector identities:

$$\nabla \cdot \nabla \times \mathbf{A} = 0, \tag{2.2.1}$$

† The added term occurs as $(1/c)\partial \mathbf{D}/\partial t$ in Maxwell's equations, hence the designation *displacement current.*

‡ In their original form, Maxwell's equations read

$$\nabla \cdot \mathbf{D} = 4\pi\rho \qquad \nabla \cdot \mathbf{B} = 0$$

$$\nabla \times \mathbf{E} + \frac{1}{c}\frac{\partial \mathbf{B}}{\partial t} = 0 \qquad \nabla \times \mathbf{H} = \frac{4\pi}{c}\mathbf{j} + \frac{1}{c}\frac{\partial \mathbf{D}}{\partial t}$$

for arbitrary vector field $\mathbf{A}(\mathbf{r}, t)$, and

$$\nabla \times \nabla\Phi = 0, \tag{2.2.2}$$

for arbitrary scalar field $\Phi(\mathbf{r}, t)$. Accordingly, every solenoidal (divergence-less) vector field can be represented as the curl of another vector field, while every irrotational vector field can be represented as the gradient of a scalar field. The second Maxwell–Lorentz equation (2.1.3) therefore implies

$$\mathbf{H} = \nabla \times \mathbf{A}, \tag{2.2.3}$$

where $\mathbf{A}$ is known as the *vector potential.* Substituting (2.2.3) into (2.1.4), we find

$$\nabla \times \left(\mathbf{E} + \frac{1}{c}\frac{\partial \mathbf{A}}{\partial t}\right) = 0. \tag{2.2.4}$$

The bracketed vector, being irrotational, must represent the gradient of a scalar field:

$$\mathbf{E} + \frac{1}{c}\frac{\partial \mathbf{A}}{\partial t} = -\nabla\Phi, \tag{2.2.5}$$

where Φ is called the *scalar potential.* In the static case, Φ, as defined, reduces to the ordinary Coulomb or electrostatic potential. The electric field is thereby given, in terms of the electromagnetic potentials, by

$$\mathbf{E} = -\nabla\Phi - \frac{1}{c}\frac{\partial \mathbf{A}}{\partial t}. \tag{2.2.6}$$

The electric and magnetic fields are thus uniquely determined—from (2.2.6) and (2.2.1), respectively—by the electromagnetic potentials $\mathbf{A}$ and Φ. The converse is not true, however. In fact, the alternative pair of potentials

$$\mathbf{A}' = \mathbf{A} + \nabla\chi, \qquad \Phi' = \Phi - \frac{1}{c}\frac{\partial \chi}{\partial t} \tag{2.2.7}$$

is easily verified to yield the same $\mathbf{E}$ and $\mathbf{H}$, $\chi(\mathbf{r}, t)$ being an arbitrary real function. The indifference of the electric and magnetic fields to the transformations (2.2.7) is termed *gauge invariance.* Any condition which, for computational convenience, restricts the form of $\mathbf{A}$ and Φ is said to define a *gauge.* Note that only the fields are physical observables; the electromagnetic potentials are, under most circumstances, mathematical artifacts—not unlike quantum-mechanical wavefunctions in this respect†.

† There are, however, quantum-mechanical effects in which the potentials themselves have physical significance. See Y. Aharonov and D. Bohm, *Phys. Rev.* **115**, 485 (1959); W. H. Furry and N. F. Ramsey, *Phys. Rev.* **118**, 623 (1960).

2.3 Wave Equations

Introduction of the electromagnetic potentials reduces the two homogeneous field equations (2.1.3) and (2.1.4) to identities. To express the source equations in terms of the potentials, apply (2.2.6) to (2.1.2). The result is

$$\nabla^2\Phi + \frac{1}{c}\frac{\partial}{\partial t}\nabla\cdot\mathbf{A} = -4\pi\rho. \tag{2.3.1}$$

Again, using (2.2.3) and (2.2.6) in (2.1.13), and the identity

$$\nabla\times(\nabla\times\mathbf{A}) = \nabla\nabla\cdot\mathbf{A} - \nabla^2\mathbf{A}, \tag{2.3.2}$$

we obtain

$$\left(\nabla^2 - \frac{1}{c^2}\frac{\partial^2}{\partial t^2}\right)\mathbf{A} - \nabla\left(\nabla\cdot\mathbf{A} + \frac{1}{c}\frac{\partial\Phi}{\partial t}\right) = -\frac{4\pi}{c}\mathbf{j}. \tag{2.3.3}$$

These source equations can be recast in a neater form by expedient choice of gauge. By imposing some condition on $\chi(\mathbf{r}, t)$, a gauge transformation can be effected from an unrestricted set of potentials $\mathbf{A}'$, Φ' to a set $\mathbf{A}$, Φ with desired analytical properties. The two classic alternatives are *Coulomb gauge* and *Lorentz gauge.*

Coulomb gauge is defined by the condition

$$\nabla\cdot\mathbf{A} = 0. \tag{2.3.4}$$

This follows in principle by making the gauge function a solution of

$$\nabla^2\chi = \nabla\cdot\mathbf{A}'. \tag{2.3.5}$$

But (2.3.4) can be assumed at the outset, without necessity of solving (2.3.5). Equation (2.3.1) accordingly reduces to

$$\nabla^2\Phi = -4\pi\rho \tag{2.3.6}$$

which is well known as Poisson's equation. The scalar potential is thereby determined by the charge distribution at each instant of time, as if the charges were at rest. Φ thus reduces to the ordinary Coulomb potential, which accounts in fact, for the designation *Coulomb gauge.* Equation (2.3.3) simplifies only slightly to

$$\left(\nabla^2 - \frac{1}{c^2}\frac{\partial^2}{\partial t^2}\right)\mathbf{A} - \frac{1}{c}\nabla\frac{\partial\Phi}{\partial t} = -\frac{4\pi}{c}\mathbf{j}. \tag{2.3.7}$$

An alternative gauge is suggested by rewriting (2.3.1) as

$$\left(\nabla^2 - \frac{1}{c^2}\frac{\partial^2}{\partial t^2}\right)\Phi + \frac{1}{c}\frac{\partial}{\partial t}\left(\nabla\cdot\mathbf{A} + \frac{1}{c}\frac{\partial\Phi}{\partial t}\right) = -4\pi\rho. \tag{2.3.8}$$

Obviously, the move now is to impose the condition

$$\nabla \cdot \mathbf{A} + \frac{1}{c}\frac{\partial \Phi}{\partial t} = 0, \tag{2.3.9}$$

which defines Lorentz gauge. In principle, the gauge function satisfies the inhomogeneous wave equation

$$\left(\nabla^2 - \frac{1}{c^2}\frac{\partial^2}{\partial t^2}\right)\chi = \nabla \cdot \mathbf{A}' + \frac{1}{c}\frac{\partial \Phi'}{\partial t}. \tag{2.3.10}$$

The source equations accordingly reduce to inhomogeneous wave equations:

$$\left(\nabla^2 - \frac{1}{c^2}\frac{\partial^2}{\partial t^2}\right)\Phi = -4\pi\rho \tag{2.3.11}$$

and

$$\left(\nabla^2 - \frac{1}{c^2}\frac{\partial^2}{\partial t^2}\right)\mathbf{A} = -\frac{4\pi}{c}\,\mathbf{j}. \tag{2.3.12}$$

This symmetry which Lorentz gauge imparts to the fundamental equations of electromagnetic theory has very profound physical significance. Its logical consequence is, in fact, the theory of relativity.

2.4 Special Relativity

The electromagnetic field equations (2.3.11) and (2.3.12) can be expressed in even more symmetrical form. Observe first the structure of the wave-equation operator written out in full:

$$\frac{\partial^2}{\partial x^2} + \frac{\partial^2}{\partial y^2} + \frac{\partial^2}{\partial z^2} - \frac{1}{c^2}\frac{\partial^2}{\partial t^2}. \tag{2.4.1}$$

This is *almost* the sum of four second partial derivatives. It can be made explicitly so by formally incorporating the time variable into a 4-dimensional cartesian system via the imaginary coordinate $x_4 = ict$. Each point in this 4-dimensional continuum (*Minkowski space*) is thereby associated with a *space–time* 4-*vector*

$$\mathbf{X} = (x_1, x_2, x_3, x_4) \equiv (x, y, z, ict) = (\mathbf{r}, ict). \tag{2.4.2}$$

The operator (2.4.1) correspondingly reduces to the 4-dimensional analog of the Laplacian:

$$\Box \equiv \sum_{\nu=1}^{4} \frac{\partial^2}{\partial x_\nu^2} = \nabla^2 - \frac{1}{c^2}\frac{\partial^2}{\partial t^2}, \tag{2.4.3}$$

which is known as the *d'Alembertian*†.

Components of 4-vectors are conventionally labelled by Greek indices $\nu, \mu, \lambda \ldots$ running from 1 to 4. Summation over a Greek index is understood if that index occurs twice in the same term (Einstein summation convention), eg,

$$a_\nu b_\nu \equiv \sum_{\nu=1}^{4} a_\nu b_\nu, \qquad \frac{\partial a_\nu}{\partial x_\nu} \equiv \sum_{\nu=1}^{4} \frac{\partial a_\nu}{\partial x_\nu}. \tag{2.4.4}$$

Carrying 4-vector formulation a step further, the continuity equation (2.1.8) can be expressed as the vanishing of a 4-divergence:

$$\frac{\partial \mathrm{j}_\nu}{\partial x_\nu} = 0 \tag{2.4.5}$$

in terms of the *current–charge 4-vector* (introduced by Poincaré):

$$\mathsf{J} = (\mathbf{j}, ic\rho). \tag{2.4.6}$$

Likewise the Lorentz gauge condition (2.3.9) takes the form

$$\frac{\partial \mathrm{A}_\nu}{\partial x_\nu} = 0 \tag{2.4.7}$$

in terms of the *electromagnetic potential 4-vector*

$$\mathsf{A} = (\mathbf{A}, i\Phi). \tag{2.4.8}$$

Making use of (2.4.3), (2.4.6) and (2.4.8), the electromagnetic field equations (2.3.11) and (2.3.12) combine to a single compact 4-vector relation:

$$\Box \mathrm{A}_\nu = -\frac{4\pi}{c}\,\mathrm{j}_\nu, \qquad \nu = 1 \ldots 4 \tag{2.4.9}$$

A most significant aspect of any vector relationship is the fact that its form is preserved under any rotation of coordinate axes, a property known as *covariance*. Rotation of axes is algebraically represented by an orthogonal transformation. Rotation in 4-space, which evidently preserves the form the electromagnetic field equations, is effected by a *Lorentz transformation*. Specifically, eqns (2.4.9) in one Lorentz frame become‡

$$\Box' \mathbf{A}'_\nu = -\frac{4\pi}{c}\,\mathbf{j}'_\nu \qquad \nu = 1 \ldots 4 \tag{2.4.10}$$

† This is alternatively written $\Box^2$ when $\Box$ is used for the 4-gradient operator $(\partial/\partial x_1, \partial/\partial x_2, \partial/\partial x_3, \partial/\partial x_4)$. The notation $\Box$ for the d'Alembertian is actually analogous to the symbol Δ (nabla) for the Laplacian.

‡ The constancy of the speed of light c in all frames has been implicitly assumed. This can be regarded as an experimental fact based on the Michaelson–Morley and other light propagation experiments.

in an alternative one under a transformation of 4-vectors according to

$$\begin{bmatrix} a_1' \\ a_2' \\ a_3' \\ a_4' \end{bmatrix} = \begin{bmatrix} L_{11}L_{12}L_{13}L_{14} \\ L_{21}L_{22}L_{23}L_{24} \\ L_{31}L_{32}L_{33}L_{34} \\ L_{41}L_{42}L_{43}L_{44} \end{bmatrix} \begin{bmatrix} a_1 \\ a_2 \\ a_3 \\ a_4 \end{bmatrix} \tag{2.4.11}$$

The elements of $\mathbb{L}$ conform to the orthogonality conditions

$$L_{\mu\nu}L_{\kappa\nu} = L_{\nu\mu}L_{\nu\kappa} = \delta_{\mu\kappa}. \tag{2.4.12}$$

Unlike 3-space orthogonal transformations, however, the matrix elements are not all real. In fact, if x', y', z', t' are to remain real, $L_{14}, L_{24}, L_{34}, L_{41}, L_{42}, L_{43}$ must be pure imaginaries.

The physical meaning of a Lorentz transformation can be demonstrated as follows. Suppose that the point in 4-space represented by x or x' refers to the origin of the primed coordinate system, say O'. Accordingly, $x_{0'} = y_{0'} = z_{0'} = 0$. The first three scalar equations contained in (2.4.11) take the explicit form

$$\begin{aligned} L_{11}x_{o'} + L_{12}y_{o'} + L_{13}z_{o'} + icL_{14}t_{o'} &= 0 \\ L_{12}x_{o'} + L_{22}y_{o'} + L_{23}z_{o'} + icL_{24}t_{o'} &= 0 \\ L_{31}x_{o'} + L_{32}y_{o'} + L_{33}z_{o'} + icL_{34}t_{o'} &= 0 \end{aligned} \tag{2.4.13}$$

Differentiating each of (2.4.13) wrt $t_{o'}$, we obtain

$$\begin{aligned} L_{11}v_x + L_{12}v_y + L_{13}v_z + icL_{14} &= 0 \\ L_{21}v_x + L_{22}v_y + L_{23}v_y + icL_{24} &= 0 \\ L_{31}v_x + L_{32}v_y + L_{33}v_z + icL_{34} &= 0 \end{aligned} \tag{2.4.14}$$

in which $\mathbf{v} = (v_x, v_y, v_z)$ is the velocity of O' as viewed in the unprimed coordinate system. (Correspondingly, to an observer in the primed system, O is moving with a velocity $-\mathbf{v}$.) Solution of eqns (2.4.14) follows by application of Cramer's rule:

$$v_x = -\frac{ic}{\Delta}\begin{vmatrix} L_{14}L_{12}L_{13} \\ L_{24}L_{22}L_{23} \\ L_{34}L_{32}L_{33} \end{vmatrix}, \qquad v_y = -\frac{ic}{\Delta}\begin{vmatrix} L_{11}L_{14}L_{13} \\ L_{21}L_{24}L_{23} \\ L_{31}L_{34}L_{33} \end{vmatrix},$$

$$v_y = -\frac{ic}{\Delta}\begin{vmatrix} L_{11}L_{12}L_{14} \\ L_{21}L_{22}L_{24} \\ L_{31}L_{32}L_{34} \end{vmatrix}, \qquad \Delta \equiv \begin{vmatrix} L_{11}L_{12}L_{13} \\ L_{21}L_{22}L_{23} \\ L_{31}L_{32}L_{33} \end{vmatrix}. \tag{2.4.15}$$

Clearly,

$$\mathbf{v} = \text{constant}, \tag{2.4.16}$$

showing that alternative Lorentz frames are, at most, moving relative to one another with constant velocity (cf. Fig. 2.1). The term *intertial frame* is generally used for coordinate systems which are, at most, in nonaccelerated motion with respect to some "rest frame".† The laws of electrodynamics are accordingly valid in every inertial frame.

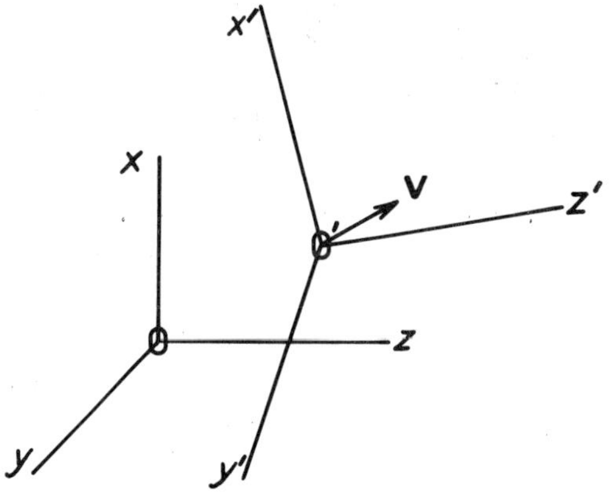

Fig. 2.1. Lorentz transformation.

Let us now explicitly construct a Lorentz transformation matrix of the simplest possible form—apart from the identity and pure rotations in 3-space. An obvious choice would have the structure

$$\mathbb{L} = \begin{bmatrix} 1 & 0 & 0 & 0 \\ 0 & 1 & 0 & 0 \\ 0 & 0 & L_{33} & L_{34} \\ 0 & 0 & L_{43} & L_{44} \end{bmatrix} \tag{2.4.17}$$

in which the 2×2 submatrix is itself orthogonal (cf. 2.4.12) with L_{34} and L_{43} being pure imaginaries. These conditions are fulfilled by

$$\mathbb{L} = \begin{bmatrix} 1 & 0 & 0 & 0 \\ 0 & 1 & 0 & 0 \\ 0 & 0 & 1/\sqrt{(1-\beta^2)} & i\beta/\sqrt{(1-\beta^2)} \\ 0 & 0 & i\beta/\sqrt{(1-\beta^2)} & 1/\sqrt{(1-\beta^2)} \end{bmatrix} \tag{2.4.18}$$

† Originally, in Newtonian mechanics, an inertial frame meant one in which Newton's laws of inertia are valid. Newton's laws do not, however, transform correctly under Lorentz transformations.

in which β is a real constant with $|\beta| < 1$.† Using the elements of (2.4.18) in (2.4.15) we find $v_x = v_y = 0$, $v_z = v = c\beta$. The primed frame is evidently moving in the z-direction with a constant velocity relative to the unprimed frame (cf. Fig. 2.2). Moreover, we identify

$$\beta = v/c. \tag{2.4.19}$$

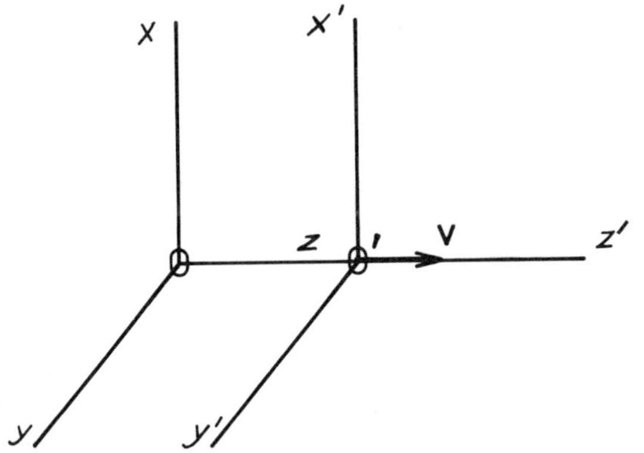

Fig. 2.2. Pure Lorentz transformation.

Explicitly, the space and time coordinates are related as follows:

$$x' = x, \quad y' = y, \quad z' = \frac{z - vt}{\sqrt{(1 - (v^2/c^2))}}, \quad t' = \frac{t - (vz/c^2)}{\sqrt{(1 - (v^2/c^2))}}. \tag{2.4.20}$$

For uniform motion in an arbitrary direction, these relations can be generalized to

$$\mathbf{r}'_\perp = \mathbf{r}_\perp, \qquad \mathbf{r}'_\parallel = \frac{\mathbf{r}_\parallel - \mathbf{v}t}{\sqrt{(1 - (v^2/c^2))}}, \qquad t' = \frac{t - (\mathbf{v}\cdot\mathbf{r}/c^2)}{\sqrt{(1 - (v^2/c^2))}} \tag{2.4.21}$$

where $\mathbf{r}_\parallel$ and $\mathbf{r}_\perp$ are, respectively, the components of the position vector parallel and transverse to the velocity $\mathbf{v}$.‡

Einstein was the first to recognize the profound implications of the Lorentz transformation as regards the intrinsic nature of space and time. Thus was born the theory of relativity, which has revolutionized our whole conception

† We have chosen the matrix elements such that det $\mathbb{L} = 1$, corresponding to a *proper* Lorentz transformation. The other possible choice, det $\mathbb{L} = -1$, represents an *improper* Lorentz transformation, which involves reflections or inversions.

‡ In the nonrelativistic limit, represented by $c \to \infty$ or $\beta \to 0$, (2.4.21) reduces to a *galilean transformation*: $\mathbf{r}'_\perp = \mathbf{r}_\perp$, $\mathbf{r}'_\parallel = \mathbf{r}_\parallel - \mathbf{v}t$, $t' = t$. Newton's laws of motion transform in accordance with galilean relativity.

of the Universe. Generalizing from electrodynamics to all of physical theory, Einstein proposed two *postulates of special relativity*:

1. The laws of physics have the same form in all inertial frames.†
2. The velocity of light (*in vacuo*) is a constant, independent of any motion of the source.

These two statements suffice to determine the form of the Lorentz transformation.‡

From the Lorentz transformation equations (2.4.21) it follows that a moving object appears to undergo contraction of its linear dimension parallel to the direction of motion, viz,

$$\mathrm{d}\mathbf{r}_{\parallel} = \mathrm{d}\mathbf{r}'_{\parallel}\sqrt{(1-(v^2/c^2))} \quad \text{(Lorentz–FitzGerald contraction)} \tag{2.4.22}$$

where $\mathrm{d}\mathbf{r}'_{\parallel}$ is the corresponding "rest length". Since there is no contraction of the transverse dimensions, an element of volume transforms according to

$$\mathrm{d}V = \mathrm{d}V'\sqrt{(1-(v^2/c^2))}. \tag{2.4.23}$$

Thus, for mass or charge density,

$$\rho = \frac{\rho'}{\sqrt{(1-(v^2/c^2))}}. \tag{2.4.24}$$

The analogous phenomenon involving time intervals is expressed by the relation

$$\mathrm{d}t = \frac{\mathrm{d}t'}{\sqrt{(1-(v^2/c^2))}}. \tag{2.4.25}$$

As first understood by Einstein, there is an apparent slowing down of moving clocks, corresponding to *time dilation*. This presents a dramatic contradiction to the Newtonian conception of an absolute time scale independent of motion.

The *local time* interval $\mathrm{d}t'$ is clearly invariant to the motion of the Lorentz frame in which it is observed. In more formal terms, the quantity

$$\mathrm{d}\tau \equiv \mathrm{d}t\sqrt{(1-(v^2/c^2))} \tag{2.4.26}$$

known as *proper time* is a Lorentz invariant. This fact can alternatively be deduced by noting that

$$(\mathrm{d}\tau)^2 = -\frac{1}{c^2}\mathrm{d}x_\nu\,\mathrm{d}x_\nu \tag{2.4.27}$$

† The special role of inertial frames is removed in the general theory of relativity. By the *principle of equivalence*, every accelerated motion is locally equivalent to some gravitational field.

‡ Postulate 1 implies that the transformation relating X' to X is linear, Postulate 2, that it is orthogonal. The argument beginning with eqn (2.4.11) can then be invoked. For details, see, for example, V. Fock, "The Theory of Space Time and Gravitation", Pergamon Press, 1959, pp. 12ff.

which, being the inner product of two 4-vectors, represents a 4-scalar. A scalar is, of course, unchanged by orthogonal transformations and is thus Lorentz invariant.

Dividing a 4-vector by a scalar gives another 4-vector. One can thereby define the *velocity 4-vector*

$$u_\nu \equiv \frac{dx_\nu}{d\tau} \tag{2.4.28}$$

which plays an important role in relativistic mechanics. Explicitly,

$$\mathbf{U} = \left(\frac{\mathbf{v}}{\sqrt{(1 - (v^2/c^2))}}, \frac{ic}{\sqrt{(1 - (v^2/c^2))}} \right) \tag{2.4.29}$$

where $\mathbf{v}$ is the 3-velocity $d\mathbf{r}/dt$. The structure of $\mathbf{U}$ also follows from the current 4-vector $(\mathbf{j}, ic\rho)$ by virtue of the convection relation $\mathbf{j} = \rho\mathbf{v}$ and the density transformation formula (2.4.24). The inner product of $\mathbf{U}$ with itself is, of course, an invariant quantity, in fact,

$$u_\nu u_\nu = -c^2. \tag{2.4.30}$$

2.5 Relativistic Dynamics

All the laws of physics must conform to the principle of special relativity. The equations of Newtonian mechanics cannot therefore have general validity because they are *not* invariant under Lorentz transformations. Still, Newton's equation of motion

$$\frac{d\mathbf{p}}{dt} = \mathbf{F} \tag{2.5.1}$$

is empirically correct in the nonrelativistic domain, so that its generalization in relativistic dynamics must reduce to it in the limit $\beta \to 0$.

All forces, whatever their origin, must have the same transformation properties since an equilibrium between forces in one Lorentz frame must be preserved in alternative frames. Electromagnetic forces provide the most accessible model. In the simplest instance, a particle of charge q in an electric field experiences a force

$$\mathbf{F}_{\text{elec}} = q\mathbf{E} \tag{2.5.2}$$

which stems, in fact, from the rudimentary definition of an electric field as the force on a unit test charge.

Let us first develop the nonrelativistic theory for the motion of a particle of mass m and charge q in an electric field. Newton's equation becomes

$$\frac{d}{dt}(m\mathbf{v}) = q\mathbf{E}. \tag{2.5.3}$$

(This is, in fact, relativistically correct for a Lorentz frame in which the particle is instantaneously at rest.) In terms of the electromagnetic potentials (cf 2.2.6)

$$\mathbf{E} = -\nabla\Phi - \frac{1}{c}\frac{\partial \mathbf{A}}{\partial t}. \tag{2.5.4}$$

It is convenient to introduce the total time derivative of the vector potential, whereby

$$\frac{\mathrm{d}\mathbf{A}}{\mathrm{d}t} = \frac{\partial \mathbf{A}}{\partial t} + \sum_{i=1}^{3} \frac{\partial \mathbf{A}}{\partial x_i}\dot{x}_i = \frac{\partial \mathbf{A}}{\partial t} + (\mathbf{v}\cdot\nabla)\mathbf{A}. \tag{2.5.5}$$

We also make use of the vector identity

$$\mathbf{v} \times (\nabla \times \mathbf{A}) = \nabla(\mathbf{v}\cdot\mathbf{A}) - (\mathbf{v}\cdot\nabla)\mathbf{A} \tag{2.5.6}$$

noting that (cf 2.2.3)

$$\mathbf{H} = \nabla \times \mathbf{A}, \tag{2.5.7}$$

which, however, vanishes since no magnetic field was assumed. The equation of motion accordingly reduces to

$$\frac{\mathrm{d}}{\mathrm{d}t}\left(m\mathbf{v} + \frac{q}{c}\mathbf{A}\right) - \nabla\left(\frac{q}{c}\mathbf{v}\cdot\mathbf{A} - q\Phi\right) = 0. \tag{2.5.8}$$

This obviously has the structure of Lagrange's equations

$$\frac{\mathrm{d}}{\mathrm{d}t}\frac{\partial L}{\partial v_i} - \frac{\partial L}{\partial x_i} = 0, \qquad i = 1, 2, 3 \tag{2.5.9}$$

with the Lagrangian function

$$L(\mathbf{r}, \mathbf{v}, t) = \tfrac{1}{2}mv^2 + \frac{q}{c}\mathbf{v}\cdot\mathbf{A} - q\Phi. \tag{2.5.10}$$

The Lagrangian (2.5.10) transforms as a scalar in 3-space. Its relativistic generalization must analogously have the form of a scalar in 4-space. By replacing the 3-vector **v** by the 4-vector **U** we arrive at a covariant Lagrangian

$$L(\mathbf{X}, \mathbf{U}) = \frac{m}{2}u_\nu u_\nu + \frac{q}{c}u_\nu A_\nu. \tag{2.5.11}$$

We postulate that this does indeed represent the relativistic dynamics of a particle in an electromagnetic field.† Correspondingly, Lagrange's equa-

† The second part represents specifically the interaction between a particle and an electromagnetic field:

$$L_{\text{int}} = \frac{q}{c}u_\nu A_\nu.$$

For continuous charge distributions, the corresponding *Lagrangian density* is given by

$$\mathscr{L}_{\text{int}} = \frac{1}{c}j_\nu A_\nu.$$

This is, in fact, the simplest nontrivial way in which two vector fields can interact, consistent with the requirements of special relativity.

tions must now be replaced by their covariant analogues

$$\frac{\mathrm{d}}{\mathrm{d}\tau}\frac{\partial L}{\partial u_\mu} - \frac{\partial L}{\partial x_\mu} = 0, \qquad \mu = 1\ldots 4. \tag{2.5.12}$$

Let us now reverse the steps in the derivation to get back to the relativistic form of Newton's law. Putting (2.5.11) into (2.5.12), we find

$$\frac{\mathrm{d}}{\mathrm{d}\tau}(mu_\mu) = \frac{q}{c}\left(\frac{\partial}{\partial x_\mu} u_\nu A_\nu - \frac{\mathrm{d}A_\mu}{\mathrm{d}\tau}\right). \tag{2.5.13}$$

The first three components can be represented by the vector equation

$$\frac{\mathrm{d}}{\mathrm{d}t}\left(\frac{m\mathbf{v}}{\sqrt{(1-(v^2/c^2))}}\right) = \frac{q}{c}\left[\nabla(\mathbf{v}\cdot\mathbf{A} - c\Phi) - \frac{\mathrm{d}\mathbf{A}}{\mathrm{d}t}\right] \tag{2.5.14}$$

having made use of (2.4.26) and (2.4.29) and multiplied through by $\sqrt{(1-(v^2/c^2))}$. Using (2.5.4)–(2.5.7) to get back the electric and magnetic fields, we arrive finally at the relativistic generalization of (2.5.3):

$$\frac{\mathrm{d}}{\mathrm{d}t}\left(\frac{m\mathbf{v}}{\sqrt{(1-(v^2/c^2))}}\right) = q\left(\mathbf{E} + \frac{1}{c}\mathbf{v}\times\mathbf{H}\right). \tag{2.5.15}$$

On the right-hand side we have no longer assumed that $\nabla\times\mathbf{A} = 0$. In view of the transformation properties of the 4-gradient and the 4-potential, vanishing of $\mathbf{H}$ in the instantaneous rest frame does *not* imply its vanishing in other Lorentz frames. Thus a field which in one frame is purely electric will appear in a relatively moving frame to have a magnetic component as well.† Evidently, a magnetic field exerts a force on a moving charge given by

$$\mathbf{F}_{\mathrm{mag}} = \frac{q}{c}\mathbf{v}\times\mathbf{H}. \tag{2.5.16}$$

The sum of (2.5.2) and (2.5.16) is called the *Lorentz force:*

$$\mathbf{F} = q\mathbf{E} + \frac{q}{c}\mathbf{v}\times\mathbf{H}. \tag{2.5.17}$$

Since all forces must transform in the same way, the fundamental equation of relativistic dynamics must have the form

$$\frac{\mathrm{d}}{\mathrm{d}t}\left(\frac{m\mathbf{v}}{\sqrt{(1-(v^2/c^2))}}\right) = \mathbf{F}. \tag{2.5.18}$$

† Specifically, the fields transform as follows:

$$\mathbf{E}'_{\parallel} = \mathbf{E}_{\parallel}, \qquad \mathbf{E}'_{\perp} = \left[\mathbf{E} + \frac{1}{c}\mathbf{v}\times\mathbf{H}\right]_{\perp}\Big/\sqrt{(1-(v^2/c^2))}$$

$$\mathbf{H}'_{\parallel} = \mathbf{H}_{\parallel}, \qquad \mathbf{H}'_{\perp} = \left[\mathbf{H} - \frac{1}{c}\mathbf{v}\times\mathbf{E}\right]_{\perp}\Big/\sqrt{(1-(v^2/c^2))}$$

Since this retains the structure of Newton's equation (2.5.1), one can identify as the relativistic momentum

$$\mathbf{p} = \frac{m\mathbf{v}}{\sqrt{(1 - (v^2/c^2))}} \tag{2.5.19}$$

which reduces, of course, to the nonrelativistic form $\mathbf{p} = m\mathbf{v}$ as $\beta \to 0$.

To determine the relativistic analogue of kinetic energy, one can proceed, just as in the nonrelativistic case, by scalar multiplication of the force equation by the particle velocity:

$$\mathbf{v} \cdot \frac{\mathrm{d}}{\mathrm{d}t}\left(\frac{m\mathbf{v}}{\sqrt{(1 - (v^2/c^2))}}\right) = \mathbf{v} \cdot \mathbf{F}. \tag{2.5.20}$$

The right-hand side represents the rate at which the force does work on the particle, hence the rate of increase of the particle's kinetic energy. By transforming the left-hand side to a time derivative, we find†

$$\frac{\mathrm{d}}{\mathrm{d}t}\left(\frac{mc^2}{\sqrt{(1 - (v^2/c^2))}}\right) = \frac{\mathrm{d}T}{\mathrm{d}t}. \tag{2.5.21}$$

The parenthesis thus represents an energy quantity‡

$$E = \frac{mc^2}{\sqrt{(1 - (v^2/c^2))}} \tag{2.5.22}$$

differing from T by, at most, an additive constant. The significance of (2.5.22) becomes apparent upon expansion of the square root:

$$E = mc^2 + \tfrac{1}{2}mv^2 + \frac{3}{8}\frac{mv^4}{c^2} + \dots \tag{2.5.23}$$

The second term is the familiar Newtonian kinetic energy, succeeding

† This can be derived alternatively from the fourth component on eqn (2.5.13). In analogy with (2.5.14) and (2.5.15), it follows that

$$\frac{\mathrm{d}}{\mathrm{d}t}\left(\frac{mc}{\sqrt{(1 - (v^2/c^2))}}\right) = -\frac{q}{c}\left[\frac{1}{c}\frac{\partial}{\partial t}(\mathbf{v} \cdot \mathbf{A} - c\Phi) + \frac{\mathrm{d}\Phi}{\mathrm{d}t}\right] = \frac{q}{c}\mathbf{v} \cdot \mathbf{E} = \frac{1}{c}\mathbf{v} \cdot \mathbf{F}$$

where $\mathbf{F}$ is the Lorentz force (2.5.17). In the last step, it has been noted that the magnetic force (2.5.16) is normal to $\mathbf{v}$.

‡ We have regarded the mass m as an invariant quantity. An alternative viewpoint introduces a velocity-dependent mass

$$m = m_0/\sqrt{(1 - (v^2/c^2))}$$

where m_0 (our m) represents the rest mass. The Newtonian form $\mathbf{p} = m\mathbf{v}$ is thereby preserved for the relativistic momentum (2.5.19) while the relativistic energy (2.5.22) is given by the famous formula $E = mc^2$.

terms are evidently relativistic corrections to the kinetic energy. The first term, however, is uniquely relativistic. It represents the *rest energy*

$$E_0 = mc^2 \tag{2.5.24}$$

which a particle possesses solely by virtue of its mass. A fundamental equivalence of matter and energy is thus manifested, allowing of their interconversion in accordance with

$$\Delta E = \Delta mc^2 \tag{2.5.25}$$

Einstein's famous relationship has, of course, been amply verified in nuclear reactions and, more spectacularly, in high energy processes accompanied by creation or annihilation of material particles.

Comparing the structure of (2.5.19) with (2.4.29), it follows that relativistic momentum can be formulated as a 4-vector proportional to 4-velocity:

$$p_\nu = mu_\nu, \tag{2.5.26}$$

so that

$$\mathbf{P} = \left(\frac{m\mathbf{v}}{\sqrt{(1-(v^2/c^2))}}, \frac{imc}{\sqrt{(1-(v^2/c^2))}}\right). \tag{2.5.27}$$

For the case of a free particle (no potential energy), the fourth component can be identified with the energy (2.5.22) according to

$$p_4 = \frac{iE}{c}. \tag{2.5.28}$$

Since momentum and energy are tied together in the same 4-vector, the separate conservation laws for momentum, energy and mass emerge as facets of a single unified principle.

The inner product of the momentum 4-vector with itself gives

$$p_\nu p_\nu = -mc^2 \tag{2.5.29}$$

This leads to the relativistic energy–momentum relation

$$E^2 = p^2c^2 + m^2c^4. \tag{2.5.30}$$

This follows alternatively by eliminating $\mathbf{v}$ between (2.5.19) and (2.5.22). Setting $E = mc^2 + T$ and passing to the limit $c \to \infty$, (2.5.30) reduces to the familiar nonrelativistic kinetic energy-momentum relationship

$$T = \frac{p^2}{2m}. \tag{2.5.31}$$

2.6 Dynamics of Electromagnetic Interactions

In the preceding section, the motion of a particle in an electromagnetic field served as a prototype for the structure of relativistic dynamics. Let us return now to consider this fundamental interaction between matter and radiation. It is of interest to develop the covariant Hamiltonian formulation analogous to (2.5.11) and (2.5.12). First, the canonical momenta associated with the Lagrangian (2.5.11) are given by

$$p_\nu = \frac{\partial L}{\partial u_\nu} = mu_\nu + \frac{q}{c} A_\nu. \tag{2.6.1}$$

The 3-vector components correspond to

$$\mathbf{p} = \frac{m\mathbf{v}}{\sqrt{(1 - (v^2/c^2))}} + \frac{q}{c}\mathbf{A} \tag{2.6.2}$$

while

$$p_4 = \frac{i}{c}\left(\frac{mc^2}{\sqrt{(1 - (v^2/c^2))}} + q\Phi\right). \tag{2.6.3}$$

Just as in the free-particle case (cf 2.5.28),

$$p_4 = \frac{iE}{c} \tag{2.6.4}$$

where the energy E now includes the electrostatic potential $q\Phi$.†

According to (2.6.1), an electromagnetic field (or any other velocity dependent field) contributes an additional term to the canonical momentum of a particle. The quantity p_ν is sometimes designated *dynamical momentum*, to distinguish it from the ordinary *kinematical momentum*

$$\pi_\nu \equiv mu_\nu = p_\nu - \frac{q}{c} A_\nu. \tag{2.6.5}$$

Note also that

$$\pi_4 = \frac{i}{c}(E - q\Phi). \tag{2.6.6}$$

A covariant Hamiltonian should now result from the 4-variable Legendre transformation

$$H(x, p) = p_\nu u_\nu - L(x, u). \tag{2.6.7}$$

Eliminating dependence on u_ν using (2.6.1) and collecting terms we find

$$H = \frac{1}{2m}\left(p_\nu - \frac{q}{c}A_\nu\right)\left(p_\nu - \frac{q}{c}A_\nu\right). \tag{2.6.8}$$

† There is evidently no contribution from the vector potential. This arises from the fact that magnetic forces do no work on a particle since they act only normal to its velocity.

Noting, however, that

$$H = \frac{\pi_\nu \pi_\nu}{2m} = \frac{m}{2} u_\nu u_\nu \tag{2.6.9}$$

and using (2.4.30), we conclude that the covariant Hamiltonian has the constant value

$$H = -\frac{mc^2}{2}. \tag{2.6.10}$$

Equations (2.6.8) and (2.6.10) combine to give the important invariance relation†

$$\left(p_\nu - \frac{q}{c}A_\nu\right)\left(p_\nu - \frac{q}{c}A_\nu\right) + m^2c^2 = 0. \tag{2.6.11}$$

More explicitly, in 3-vector notation

$$(E - q\Phi)^2 = \left(\mathbf{p} - \frac{q}{c}\mathbf{A}\right)^2 c^2 + m^2c^4, \tag{2.6.12}$$

which is the generalization of the free-particle energy-momentum relation (2.5.30). This reduces in the nonrelativistic limit, in analogy with (2.5.31), to

$$E = \frac{1}{2m}\left(\mathbf{p} - \frac{q}{c}\mathbf{A}\right)^2 + q\Phi, \tag{2.6.13}$$

excluding the rest energy mc^2.

The covariant Hamiltonian is not otherwise a particularly useful dynamical variable. It cannot, for example, be identified with an energy: E, being a component of a 4-vector, is not itself a covariant quantity. For our subsequent development of the quantum theory of electromagnetic interactions, we shall require the *nonrelativistic* Hamiltonian formulation. This can be based upon Newton's law for a particle acted upon by the Lorentz force (2.5.17):

$$\frac{\mathrm{d}}{\mathrm{d}t}(m\mathbf{v}) = q\mathbf{E} + \frac{q}{c}\mathbf{v} \times \mathbf{H} \tag{2.6.14}$$

or the equivalent Lagrangian function (cf 2.5.10)

$$L(\mathbf{r}, \mathbf{v}, t) = \tfrac{1}{2}mv^2 + \frac{q}{c}\mathbf{v} \cdot \mathbf{A} - q\Phi. \tag{2.6.15}$$

† The free-particle momentum relation (2.5.29) pertains as well to the kinetic momentum for a particle in a field, viz,

$$\pi_\nu \pi_\nu + m^2c^2 = 0.$$

Thus, by use of (2.6.5) and (2.6.6), we arrive alternatively at eqns (2.6.11)–(2.6.13).

The generalized momenta are determined by

$$p_i = \frac{\partial L}{\partial v_i} = mv_i + \frac{q}{c}A_i, \qquad i = 1, 2, 3 \tag{2.6.16}$$

being the nonrelativistic limit of (2.6.2). In vector form, the velocity and momentum of a nonrelativistic particle in an electromagnetic field are related by

$$\mathbf{p} = m\mathbf{v} + \frac{q}{c}\mathbf{A}. \tag{2.6.17}$$

The nonrelativistic Hamiltonian function can now be constructed using

$$H = \mathbf{p} \cdot \mathbf{v} - L. \tag{2.6.18}$$

The result is

$$H = \frac{1}{2m}\left(\mathbf{p} - \frac{q}{c}\mathbf{A}\right)^2 + q\Phi. \tag{2.6.19}$$

Comparison with (2.6.13) shows that this Hamiltonian can be identified with the energy of a particle in an electromagnetic field. The Lagrangian (2.6.15) has indeed the requisite structure, in accord with (1.2.15)–(1.2.17).

A. *Electric and Magnetic Moments*

A localized distribution of charges and currents can be characterized by a discrete set of constants called monopoles, dipoles, quadrupoles, octupoles, etc. Let us first define the electric moments by considering a continuous charge distribution $\rho(\mathbf{r})$ in an electrostatic field $\Phi = \Phi(\mathbf{r})$, $\mathbf{A} = 0$. The Hamiltonian has the form

$$H = H^{(0)} + \int \rho(\mathbf{r})\,\Phi(\mathbf{r})\,\mathrm{d}^3\mathbf{r} \tag{2.6.20}$$

$H^{(0)}$ denoting its field-free part. Suppose now that $\rho(\mathbf{r})$ is predominantly localized in the region about $\mathbf{r} = 0$. It is appropriate then to develop the potential in a Taylor series

$$\Phi(\mathbf{r}) = \Phi_0 + \mathbf{r} \cdot \nabla_0 \Phi_0 + \tfrac{1}{2}(\mathbf{r} \cdot \nabla_0)^2 \Phi_0 + \ldots \tag{2.6.21}$$

The subscript "0" is used to denote the limit $\mathbf{r} = 0$. By introducing the electric field at the origin

$$\mathbf{E}_0 = -\nabla_0 \Phi_0 \tag{2.6.22}$$

the expansion can be written

$$\Phi(\mathbf{r}) = \Phi_0 - \mathbf{r} \cdot \mathbf{E}_0 - \tfrac{1}{2}\mathbf{r}\mathbf{r}{:}\nabla_0 \mathbf{E}_0 - \ldots \tag{2.6.23}$$

The last term, written as a double scalar product, means

$$-\tfrac{1}{2}\sum_{i,j=1}^{3} x_i x_j \left(\frac{\partial E_i}{\partial x_j}\right)_0 \qquad (2.6.24)$$

Substitution of (2.6.23) into (2.6.20) now results in the following series for the Hamiltonian:

$$H = H^{(0)} + q\Phi_0 - \boldsymbol{\mu}\cdot\mathbf{E}_0 - \tfrac{1}{2}\mathfrak{Q}{:}\nabla_0\mathbf{E}_0 \qquad (2.6.25)$$

where

$$q \equiv \int \rho(\mathbf{r})\,\mathrm{d}^3\mathbf{r} \qquad (2.6.26)$$

the total electric charge or "monopole moment";

$$\boldsymbol{\mu} \equiv \int \mathbf{r}\rho(\mathbf{r})\,\mathrm{d}^3\mathbf{r} \qquad (2.6.27)$$

the electric dipole moment (about the origin $\mathbf{r} = 0$) and

$$\mathfrak{Q}' \equiv \int \mathbf{r}\mathbf{r}\rho(\mathbf{r})\,\mathrm{d}^3\mathbf{r} \qquad (2.6.28)$$

the electric quadrupole moment tensor. Evidently, each multipole couples with one aspect of the external field: the monopole with the potential, the dipole with the field, the quadrupole with the field gradient, etc.

For a spherically-symmetrical distribution of charge, $\rho(\mathbf{r}) = \rho(r)$, all off-diagonal elements of the quadrupole vanish. In fact

$$Q'_{ij} = \int x_i x_j \rho(\mathbf{r})\,\mathrm{d}^3\mathbf{r} = \delta_{ij}\int r^2\rho(r)\,\mathrm{d}^3\mathbf{r} \qquad (2.6.29)$$

The quadrupole energy term reduces accordingly to

$$E_{\text{quad}} = -\tfrac{1}{2}\sum_{ij} Q'_{ij}\left(\frac{\partial E_i}{\partial x_j}\right)_0 = -\tfrac{1}{6}\,(\nabla_0\cdot\mathbf{E}_0)\int r^2\rho(r)\,\mathrm{d}^3\mathbf{r} = 0 \qquad (2.6.30)$$

since, by the first of Maxwell's equations, $\nabla\cdot\mathbf{E} = 0$ external to the sources of the field. For a spherical charge distribution, in fact, *all* multipole energies higher than the monopole necessarily vanish.

It has been found convenient to redefine the electric quadrupole tensor as follows:

$$Q_{ij} \equiv \int (3x_i x_j - r^2\delta_{ij})\rho(\mathbf{r})\,\mathrm{d}^3\mathbf{r} \qquad (2.6.31)$$

or, in diadic notation,

$$\mathfrak{Q} \equiv \int (3\mathbf{rr} - r^2\mathfrak{J})\rho(\mathbf{r})\,\mathrm{d}^3\mathbf{r} \tag{2.6.32}$$

in which $\mathfrak{J} \equiv \mathbf{ii} + \mathbf{jj} + \mathbf{kk}$, the unit diadic. By this definition, $\mathfrak{Q}$ is traceless, ie,

$$\mathrm{tr}\,\mathfrak{Q} \equiv \sum_i Q_{ii} = 0. \tag{2.6.33}$$

Moreover $\mathfrak{Q} = 0$ now for a spherical charge distribution. The quadrupole energy term in (2.6.25) is now written

$$E_{\text{quad}} = -\tfrac{1}{6}\mathfrak{Q}:\nabla_0\mathbf{E}_0. \tag{2.6.34}$$

This reduces to a particularly simple form for a cylindrically-symmetrical charge distribution. Let z be the axis of symmetry. Now

$$Q_{zz} = \int (3z^2 - r^2)\rho(\mathbf{r})\,\mathrm{d}^3\mathbf{r} = 2\int r^2 P_2(\cos\theta)\rho(\mathbf{r})\,\mathrm{d}^3\mathbf{r} \tag{2.6.35}$$

in which† $P_2(\cos) = \frac{1}{2}(3\cos^2\theta - 1)$, the second Legendre polynomial. Also

$$Q_{xx} = Q_{yy} = -\tfrac{1}{2}Q_{zz} \tag{2.6.36}$$

by virtue of the cylindrical symmetry and vanishing trace. All off-diagonal elements vanish. Therefore

$$E_{\text{quad}} = -\frac{1}{6}\left[Q_{xx}\left(\frac{\partial E_x}{\partial x}\right)_0 + Q_{yy}\left(\frac{\partial E_y}{\partial y}\right)_0 + Q_{zz}\left(\frac{\partial E_z}{\partial z}\right)_0\right]. \tag{2.6.37}$$

But by (2.6.36) and the vanishing of the divergence, viz,

$$\left(\frac{\partial E_x}{\partial x}\right)_0 + \left(\frac{\partial E_y}{\partial y}\right)_0 = -\left(\frac{\partial E_z}{\partial z}\right)_0 \tag{2.6.38}$$

we have

$$E_{\text{quad}} = -\tfrac{1}{4}Q\left(\frac{\partial E_z}{\partial z}\right)_0 \tag{2.6.39}$$

writing $Q \equiv Q_{zz}$ as *the* quadrupole moment. (In nuclear physics, this quantity is usually written eQ.) For positive charges, $Q > 0$ corresponds to a *prolate* shape—elongation along the symmetry axis with $\langle z^2\rangle > \frac{1}{3}\langle r^2\rangle$—while $Q < 0$ corresponds to an *oblate* shape.

Since the quadrupole tensor is symmetric, it can always be brought into

† Multipole moment elements can be defined more systematically in terms of spherical harmonics as follows:

$$q_{lm} \equiv \int r^l Y^*_{lm}(\theta, \psi)\rho(\mathbf{r})\,\mathrm{d}^3\mathbf{r}$$

diagonal form by an appropriate choice of principal axes. Thus, even for an asymmetric charge distribution, one can find a representation

$$\mathfrak{Q} = \mathbf{ii}Q_1 + \mathbf{jj}Q_2 + \mathbf{kk}Q_3 . \tag{2.6.40}$$

By virtue of the vanishing trace, only two of the three principal moments are independent.

Magnetic interactions can likewise be parametrized in terms of magnetic multipole moments. We shall require only the magnetic dipole moment. It is sufficient therefore to consider a constant homogeneous magnetic field $\mathbf{H}_0$. This is conveniently represented by the vector potential

$$\mathbf{A}(\mathbf{r}) = \tfrac{1}{2}\mathbf{H}_0 \times \mathbf{r} . \tag{2.6.41}$$

Putting $\mathbf{A}(\mathbf{r})$ into the one-particle Hamiltonian function (2.6.19) we obtain

$$\begin{aligned} H &= \frac{1}{2m}\left(\mathbf{p} - \frac{q}{2c}\mathbf{H}_0 \times \mathbf{r}\right)^2 \\ &= \frac{p^2}{2m} - \frac{q}{2mc}\mathbf{p}\cdot\mathbf{H}_0 \times \mathbf{r} + \frac{q^2}{8mc^2}(\mathbf{H}_0 \times \mathbf{r})^2 . \end{aligned} \tag{2.6.42}$$

The last term accounts for diamagnetism. By rearranging the term linear in $\mathbf{H}_0$, we obtain

$$-\frac{q}{2mc}\mathbf{l}\cdot\mathbf{H}_0 \tag{2.6.43}$$

where $\mathbf{l} = \mathbf{r} \times \mathbf{p}$ is the angular momentum of the particle. One can write therefore, in analogy with the electric dipole term in (2.6.25):

$$E_{\text{mag dip}} = -\,\mathbf{m}\cdot\mathbf{H}_0 \tag{2.6.44}$$

in terms of the magnetic dipole moment

$$\mathbf{m} \equiv \frac{q}{2mc}\mathbf{l} . \tag{2.6.45}$$

The ratio between magnetic moment and angular momentum is called the gyromagnetic (or magnetigyric) ratio, denoted γ. For orbital motion in both classical and quantum mechanics

$$\gamma = q/2mc . \tag{2.6.46}$$

For electron spin, however, ($q = -e, \mathbf{l} = \mathbf{s}$)

$$\gamma = -ge/2mc \tag{2.6.47}$$

with $g = 2{\cdot}0023$. Nuclear-spin g-factors are even more anomalous.

For continuous charge–current distributions, one can generalize the definition of magnetic dipole moment to

$$\mathbf{m} \equiv \frac{1}{2c}\int \mathbf{r} \times \mathbf{v}\rho(\mathbf{r})\,\mathrm{d}^3\mathbf{r} = \frac{1}{2c}\int \mathbf{r} \times \mathbf{j}(\mathbf{r})\,\mathrm{d}^3\mathbf{r}. \tag{2.6.48}$$

Electric and magnetic dipole moments are sometimes defined in terms of derivatives of the Hamiltonian with respect to field components, viz,

$$\boldsymbol{\mu} \equiv -\nabla_{\mathbf{E}} H|_{\mathbf{E}=0} \quad \text{and} \quad \mathbf{m} \equiv -\nabla_{\mathbf{H}} H|_{\mathbf{H}=0}. \tag{2.6.49}$$

These relations are easily inferred from eqns (2.6.25) and (2.6.42), respectively.

2.7 Energy of an Electromagnetic Field

The mechanical energy imparted to a particle by an electromagnetic field was calculated in Section 2.5 (cf. eqn 2.5.20 ff). Specifically,

$$\frac{\mathrm{d}E_{\text{mech}}}{\mathrm{d}t} = \mathbf{v}\cdot\mathbf{F} = q\mathbf{v}\cdot\mathbf{E} \tag{2.7.1}$$

where **F** is the Lorentz force. If the energy conservation principle (or its relativistic analog) is assumed, the field itself must have lost an equal amount of energy, ie,

$$\frac{\mathrm{d}E_{\text{field}}}{\mathrm{d}t} = -\frac{\mathrm{d}E_{\text{mech}}}{\mathrm{d}t} = -q\mathbf{v}\cdot\mathbf{E}. \tag{2.7.2}$$

It is convenient to introduce $\mathscr{U}$ the *energy density* of the field, defined such that

$$\int \mathscr{U}\,\mathrm{d}\tau = E_{\text{field}}. \tag{2.7.3}$$

We can then write, for a continuous distribution of charge and current,

$$\frac{\mathrm{d}\mathscr{U}}{\mathrm{d}t} = -\mathbf{j}\cdot\mathbf{E}. \tag{2.7.4}$$

All reference to the material system can now be removed by substituting for **j** from the fourth Maxwell–Lorentz equation (2.1.13):

$$\frac{\mathrm{d}\mathscr{U}}{\mathrm{d}t} = -\frac{c}{4\pi}\mathbf{E}\cdot\nabla\times\mathbf{H} + \frac{1}{4\pi}\mathbf{E}\cdot\frac{\partial\mathbf{E}}{\partial t}. \tag{2.7.5}$$

Noting the vector identity

$$\nabla(\mathbf{E}\times\mathbf{H}) = \mathbf{H}\cdot\nabla\times\mathbf{E} - \mathbf{E}\cdot\nabla\times\mathbf{H}, \tag{2.7.6}$$

and substituting for $\nabla \times \mathbf{E}$ from (2.1.4), we obtain

$$\frac{d\mathscr{U}}{dt} = \frac{1}{4\pi}\left(\mathbf{E}\cdot\frac{\partial \mathbf{E}}{\partial t} + \mathbf{H}\cdot\frac{\partial \mathbf{H}}{\partial t}\right) + \frac{c}{4\pi}\nabla\cdot(\mathbf{E}\times\mathbf{H}). \tag{2.7.7}$$

Rearranging,

$$\frac{d\mathscr{U}}{dt} = \frac{\partial}{\partial t}\left[\frac{1}{8\pi}(E^2 + H^2)\right] + \nabla\cdot\left[\frac{c}{4\pi}\mathbf{E}\times\mathbf{H}\right], \tag{2.7.8}$$

which can be interpreted as an equation of continuity for electromagnetic field energy, with a source term representing the net rate of conversion from mechanical energy (cf 2.1.9). The energy density of an electromagnetic field in free space is thus†

$$\mathscr{U} = \frac{1}{8\pi}(E^2 + H^2). \tag{2.7.9}$$

The energy flux, known as *Poynting's vector*, is correspondingly

$$\mathbf{S} = \frac{c}{4\pi}\mathbf{E}\times\mathbf{H}. \tag{2.7.10}$$

This quantity has dimensions of energy flux across unit area per unit time.

The momentum density $\mathbf{G}$ of the field can also be calculated by an analogous argument, starting with $d\mathbf{p}_{mech}/dt = \mathbf{F}$ and assuming conservation of total momentum of particle and field. The result is

$$\mathbf{G} = \frac{1}{4\pi c}\mathbf{E}\times\mathbf{H} = \frac{1}{c^2}\mathbf{S}. \tag{2.7.11}$$

2.8 Radiation Fields

A fundamental feature of electromagnetic theory is the existence of wavelike solutions to the field equations representing coupled electric and magnetic fields propagating through space at a speed c and capable of transporting energy and information. Following the method of Hertz, we shall obtain general expressions for the fields produced by a localized distribution of charges and currents. The calculation is simplified by relating the charge and current densities to a single source function $\mathbf{P}(\mathbf{r}, t)$, defined such that

$$\rho = -\nabla\cdot\mathbf{P} \tag{2.8.1}$$

and

$$\mathbf{j} = \partial\mathbf{P}/\partial t. \tag{2.8.2}$$

† This result can also be deduced from electrical circuit theory. The energy of a charged capacitor, $\frac{1}{2}Cq^2$, can be equated to a volume integral over the electrical field it produces: $\int (E^2/8\pi)\,d\tau$. Correspondingly, the energy of a current-carrying inductor, $\frac{1}{2}Li^2$, is equal to $\int (H^2/8\pi)\,d\tau$.

The equation of continuity (2.1.8) is thereby reduced to an identity. Analogously, the scalar and vector potentials can be related to the *Hertzian vector* or *superpotential* $\mathbf{\Pi}(\mathbf{r}, t)$ by means of

$$\Phi = -\nabla\cdot\mathbf{\Pi}, \qquad \mathbf{A} = \frac{1}{c}\frac{\partial\mathbf{\Pi}}{\partial t}. \tag{2.8.3}$$

The Lorentz condition (2.3.9) is, in this way, identically satisfied. In terms of the source function and superpotential, the wave equations (2.3.11) and (2.3.12) combine to

$$\Box\mathbf{\Pi} = -4\pi\mathbf{P} \tag{2.8.4}$$

As shown in Section 6.1D, eqn (6.1.60), this inhomogeneous wave equation has the solution

$$\mathbf{\Pi}(\mathbf{r}, t) = \int \frac{[\mathbf{P}(\mathbf{r}', t')]}{R}\, \mathrm{d}^3\mathbf{r}' \tag{2.8.5}$$

where

$$[\mathbf{P}(\mathbf{r}', t')] \equiv \mathbf{P}(\mathbf{r}', t - R/c), \tag{2.8.6}$$

in terms of

$$\mathbf{R} \equiv \mathbf{r} - \mathbf{r}', \qquad R = |\mathbf{r} - \mathbf{r}'|. \tag{2.8.7}$$

It is computationally convenient to express (2.8.5) in terms of the retarded Green's function (6.1.55):

$$\mathbf{\Pi}(\mathbf{r}, t) = \iint \frac{\mathbf{P}(\mathbf{r}', t')}{R}\, \delta\left(t - t' - \frac{R}{c}\right) \mathrm{d}^3\mathbf{r}'\, \mathrm{d}t' \tag{2.8.8}$$

The electric and magnetic fields are given in terms of the superpotential by substituting (2.8.3) into (2.2.3) and (2.2.6). The results are

$$\mathbf{H} = \frac{1}{c}\frac{\partial}{\partial t}\nabla\times\mathbf{\Pi} \tag{2.8.9}$$

and

$$\mathbf{E} = \nabla\nabla\cdot\mathbf{\Pi} - \frac{1}{c^2}\frac{\partial^2\mathbf{\Pi}}{\partial t^2} \tag{2.8.10}$$

To calculate the electric field we find first

$$\frac{1}{c^2}\frac{\partial^2\mathbf{\Pi}}{\partial t^2} = \frac{1}{c^2}\int \frac{[\partial^2\mathbf{P}/\partial t^2]}{R}\, \mathrm{d}^3\mathbf{r}'. \tag{2.8.11}$$

Keeping only those terms which contribute in the radiation zone, by neglecting derivatives of $1/R$ inside the integral, we obtain

$$\nabla\nabla\cdot\mathbf{\Pi} = \frac{1}{c^2}\int\frac{[\partial^2\mathbf{P}/\partial t^2]\cdot\mathbf{RR}}{R^3}\,\mathrm{d}^3\mathbf{r}'. \tag{2.8.12}$$

Combining (2.8.11) with (2.8.12), in accordance with (2.8.10), the electric radiation field can be expressed

$$\mathbf{E}_{\text{rad}} = \frac{1}{c^2}\int\frac{([\partial^2\mathbf{P}/\partial t^2]\times\mathbf{R})\times\mathbf{R}}{R^3}\,\mathrm{d}^3\mathbf{r}'. \tag{2.8.13}$$

The dependence of (2.8.12) and (2.8.13) on $\partial^2\mathbf{P}/\partial t^2$ shows that electromagnetic radiation is produced by acceleration of charge. Harmonic oscillation of charge at a given frequency will moreover produce radiation of the same frequency.

It is convenient to write, in (2.8.12) and (2.8.13),

$$\mathbf{R} = R\mathbf{n} \tag{2.8.14}$$

where $\mathbf{n}$ represents the unit vector in the direction of propagation—from source to field point. Accordingly, making use of (2.8.11), the fields can be written

$$\mathbf{H}_{\text{rad}} = \frac{1}{c^2}\frac{\partial^2\mathbf{\Pi}}{\partial t^2}\times\mathbf{n}. \tag{2.8.15}$$

In computing the electric and magnetic fields, note that derivatives of field quantities act only on unprimed variables. We find first

$$\begin{aligned}\nabla\times\mathbf{\Pi} &= -\iint\mathbf{P}(\mathbf{r}',t')\times\nabla\frac{\delta(t-t'-(R/c))}{R}\,\mathrm{d}^3\mathbf{r}'\,\mathrm{d}t'\\ &= \iint\frac{\mathbf{P}(\mathbf{r}',t')\times\mathbf{R}}{R^3}\,\delta\left(t-t'-\frac{R}{c}\right)\mathrm{d}^3\mathbf{r}'\,\mathrm{d}t'\\ &\quad+\frac{1}{c}\iint\frac{\mathbf{P}(\mathbf{r}',t')\times\mathbf{R}}{R^2}\,\delta'\left(t-t'-\frac{R}{c}\right)\mathrm{d}^3\mathbf{r}'\,\mathrm{d}t'.\end{aligned} \tag{2.8.16}$$

Differentiating with respect to t adds a prime to each deltafunction. Using the integral properties of deltafunctions, the magnetic field (2.8.9) is

$$\mathbf{H} = \frac{1}{c}\int\frac{[\partial\mathbf{P}/\partial t]\times\mathbf{R}}{R^3}\,\mathrm{d}^3\mathbf{r}' + \frac{1}{c^2}\int\frac{[\partial^2\mathbf{P}/\partial t^2]\times\mathbf{R}}{R^2}\,\mathrm{d}^3\mathbf{r}'. \tag{2.8.17}$$

The first term, called the *induction field*,† falls off rapidly with distance

† This is the retarded magnetostatic field of a current distribution, namely

$$\frac{1}{c}\int\frac{[\mathbf{j}]\times\mathbf{R}}{R^3}\,\mathrm{d}^3\mathbf{r}'.$$

from source—as $1/R^2$. The second term in (2.8.17) has longer range, falling off as $1/R$. It represents the magnetic radiation field

$$\mathbf{H}_{\text{rad}} = \frac{1}{c^2}\int \frac{[\partial^2\mathbf{P}/\partial t^2] \times \mathbf{R}}{R^2}\, \mathrm{d}^3\mathbf{r}'. \tag{2.8.18}$$

The domain in which terms in $1/R^2$ are negligible w.r.t. those in $1/R$ is termed the *radiation zone* and

$$\mathbf{E}_{\text{rad}} = \frac{1}{c^2}\left(\frac{\partial^2 \pi}{\partial t^2} \times \mathbf{n}\right) \times \mathbf{n}. \tag{2.8.19}$$

Clearly then,

$$\mathbf{E}_{\text{rad}} = \mathbf{H}_{\text{rad}} \times \mathbf{n} \tag{2.8.20}$$

and

$$\mathbf{H}_{\text{rad}} \cdot \mathbf{n} = \mathbf{E}_{\text{rad}} \cdot \mathbf{n} = 0 \tag{2.8.21}$$

showing the electric and magnetic radiation fields to be mutually perpendicular, transverse to the direction of propagation and (in gaussian units) equal in magnitude.

2.9 Dipole Radiation

We shall now consider more explicitly the radiation emitted by an oscillating system of charges. It is convenient to represent the source function as a superposition of harmonic motions, say by a Fourier integral

$$\mathbf{P}(\mathbf{r}, t) = \int_{-\infty}^{\infty} \mathfrak{P}_{\omega}(\mathbf{r})\, \mathrm{e}^{-i\omega t}\, \mathrm{d}\omega. \tag{2.9.1}$$

The amplitudes $\mathfrak{P}_{\omega}(\mathbf{r})$ are understood to be complex quantities incorporating a phase factor:

$$\mathfrak{P}_{\omega}(\mathbf{r}) = |\mathfrak{P}_{\omega}(\mathbf{r})| \exp(i\alpha_{\omega}) \tag{2.9.2}$$

(absolute value in complex variable sense). In order that $\mathbf{P}(\mathbf{r}, t)$ represent a real function, it is necessary that the negative-frequency Fourier amplitudes satisfy the condition

$$\mathfrak{P}_{-\omega}(\mathbf{r}) = \mathfrak{P}^*_{\omega}(\mathbf{r}). \tag{2.9.3}$$

For a harmonically oscillating source (2.9.1) reduces to

$$\mathbb{P}_{\omega}(\mathbf{r}, t) = \mathfrak{P}_{\omega}(\mathbf{r})\, \mathrm{e}^{-i\omega t} + \mathfrak{P}_{-\omega}(\mathbf{r})\, \mathrm{e}^{i\omega t} \tag{2.9.4}$$

This can alternatively be written, by virtue of (2.9.2) and (2.9.3),

$$\mathbb{P}_{\omega}(\mathbf{r}, t) = 2|\mathfrak{P}_{\omega}(\mathbf{r})| \cos(\omega t + \alpha_{\omega}) \tag{2.9.5}$$

Since the exponential forms are a bit easier to manipulate we shall prefer to write

$$\mathbf{P}_\omega(\mathbf{r}, t) = \mathfrak{P}_\omega(\mathbf{r})\,\mathrm{e}^{-i\omega t} + \mathrm{cc} \tag{2.9.6}$$

(cc = complex conjugate of preceding term). The form: function + cc is preserved under all purely linear operations.

The retarded analog of (2.9.6) is given by (cf. 2.8.6):

$$\begin{aligned}[\mathbf{P}_\omega(\mathbf{r}', t')] &= \mathfrak{P}_\omega(\mathbf{r}')\exp(-i\omega(t - R/c)) + \mathrm{cc} \\ &= \mathfrak{P}_\omega(\mathbf{r}')\exp(\mathrm{i}(kR - \omega t)) + \mathrm{cc}\end{aligned} \tag{2.9.7}$$

having defined

$$k \equiv \omega/c. \tag{2.9.8}$$

By (2.8.5) this harmonic source produces a superpotential†

$$\mathbf{\Pi}_\omega(\mathbf{r}, t) = \int \frac{\mathfrak{P}_\omega(\mathbf{r}')}{R}\exp(\mathrm{i}(kR - \omega t))\,\mathrm{d}^3\mathbf{r}' + \mathrm{cc}. \tag{2.9.9}$$

This demonstrates that radiation is synchronous with its source: a harmonically oscillating source produces monochromatic radiation of the same frequency and phase.

The radiation wavelength is given by

$$\lambda = \frac{c}{\nu} = \frac{2\pi c}{\omega} = \frac{2\pi}{k}. \tag{2.9.10}$$

Now, a source of maximum dimension r' in which particles oscillate at a maximum speed v' produces radiation of a frequency approximating $\nu \sim v'/r'$. Nonrelativistic oscillation, with $v' \ll c$ implies therefore the inequality $\lambda \gg r'$ or $kr' \ll 1$. In the far radiation zone, where $r \gg \lambda$, it is sufficient to approximate‡

† If we write

$$\mathbf{\Pi}_\omega(\mathbf{r}, t) = \pi_\omega(\mathbf{r})\,\mathrm{e}^{-i\omega t} + \mathrm{cc}$$

then

$$\pi_\omega = \int \frac{\mathfrak{P}_\omega(\mathbf{r}')}{R}\,\mathrm{e}^{ikR}\,\mathrm{d}^3\mathbf{r}'$$

represents a solution of the inhomogeneous Helmholtz equation (cf. Section 6.1C)

$$(\nabla^2 + k^2)\pi_\omega = -4\pi\mathfrak{P}_\omega$$

‡ In the domain of $r \gg \lambda \gg r'$, the following expansion applies:

$$\frac{\mathrm{e}^{ikR}}{R} = \frac{\mathrm{e}^{ikr}}{r}\sum_{l=0}^{\infty}(-i)^l\frac{2^l l!}{(2l)!}(kr')^l P_l(\cos\Theta)$$

where $\cos\Theta = \mathbf{r}\cdot\mathbf{r}'/rr'$. The leading terms are explicitly

$$\frac{\mathrm{e}^{ikR}}{R} = \frac{\mathrm{e}^{ikr}}{r}\left(1 - ik\frac{\mathbf{r}\cdot\mathbf{r}'}{r} + \ldots\right).$$

$$\frac{e^{ikR}}{R} \approx \frac{e^{ikr}}{r} . \tag{2.9.11}$$

This gives rise, as we shall see, to the *dipole approximation* for the radiation field.

It is appropriate now to introduce the *Hertzian dipole vector*

$$\boldsymbol{\mu}(\omega) \equiv \int \mathfrak{P}_{\omega}(\mathbf{r})\, \mathrm{d}^3\mathbf{r} . \tag{2.9.12}$$

By virtue of (2.9.3)

$$\boldsymbol{\mu}(-\omega) = \boldsymbol{\mu}^*(\omega) . \tag{2.9.13}$$

The Hertzian dipole can be related to the charge and current densities of the source. We note first that, by partial integration (assuming boundary terms vanish),

$$\int \mathbf{r}\nabla \cdot \mathfrak{P}_{\omega}(\mathbf{r})\, \mathrm{d}^3\mathbf{r} = - \int \mathfrak{P}_{\omega}(\mathbf{r})\, \mathrm{d}^3\mathbf{r} . \tag{2.9.14}$$

Now, application of (2.8.1) on the left-hand side and (2.9.14) on the right gives

$$\boldsymbol{\mu}(\omega) = \int \mathbf{r}\rho_{\omega}(\mathbf{r})\, \mathrm{d}^3\mathbf{r} \tag{2.9.15}$$

in which $\rho_{\omega}(\mathbf{r})$ is the appropriate Fourier component of charge density. Note that (2.9.15) identifies $\boldsymbol{\mu}(\omega)$ as the centroid of a charge distribution, hence its designation as a dipole (cf. 2.6.27). Alternatively, by (2.8.2), one can write

$$\boldsymbol{\mu}(\omega) = \frac{i}{\omega} \int \mathbf{j}_{\omega}(\mathbf{r})\, \mathrm{d}^3\mathbf{r} \tag{2.9.16}$$

in terms of the Fourier component of charge density.

By applying (2.9.11) and (2.9.12) in (2.9.9) we obtain for the superpotential in the radiation zone

$$\boldsymbol{\Pi}_{\omega}(\mathbf{r}, t) = \boldsymbol{\mu}(\omega)\frac{e^{i(kr - \omega t)}}{r} + \mathrm{cc}. \tag{2.9.17}$$

Equations (2.8.15) and (2.8.19) now give the corresponding dipole radiation fields

$$\begin{aligned} \mathbf{H}_{\text{dip rad}} &= k^2\mathbf{n} \times \boldsymbol{\mu}(\omega)\frac{e^{i(kr - \omega t)}}{r} + \mathrm{cc} \\ \mathbf{E}_{\text{dip rad}} &= k^2[\mathbf{n} \times \boldsymbol{\mu}(\omega)] \times \mathbf{n}\frac{e^{i(kr - \omega t)}}{r} + \mathrm{cc} . \end{aligned} \tag{2.9.18}$$

As could be anticipated, these have the structure of outgoing spherical waves.

The radiated power can be found by applying Poynting's theorem (2.7.10). If eqns (2.9.18) are written

$$\left.\begin{aligned} \mathbf{H} &= \mathscr{H}\,\mathrm{e}^{-i\omega t} + \mathscr{H}^*\,\mathrm{e}^{i\omega t} \\ \mathbf{E} &= \mathscr{E}\,\mathrm{e}^{-i\omega t} + \mathscr{E}^*\,\mathrm{e}^{i\omega t} \\ \mathscr{H} &= k^2\mathbf{n}\times\boldsymbol{\mu}(\omega)\,\mathrm{e}^{ikr}/r, \text{ etc.} \end{aligned}\right\} \tag{2.9.19}$$

then

$$\mathbf{S} = \frac{c}{4\pi}\left[\mathscr{E}\times\mathscr{H}^* + \mathscr{E}^*\times\mathscr{H} + \mathscr{E}\times\mathscr{H}\,\mathrm{e}^{-2i\omega t} + \mathscr{E}^*\times\mathscr{H}^*\,\mathrm{e}^{2i\omega t}\right]. \tag{2.9.20}$$

One is usually interested in the flux averaged over some appropriately long time period, that is, the *intensity*:

$$\mathbf{I} \equiv \overline{\mathbf{S}} = \lim_{T\to\infty}\frac{1}{T}\int_{-T/2}^{T/2}\mathbf{S}\,\mathrm{d}t\,. \tag{2.9.21}$$

The time factors of the last two terms in (2.9.20) average to zero so that†

$$I = \frac{c}{4\pi}(\mathscr{E}\times\mathscr{H}^* + \mathscr{E}^*\times\mathscr{H}). \tag{2.9.21}$$

By (2.8.20) we have also

$$\mathscr{E} = \mathscr{H}\times\mathbf{n} \tag{2.9.22}$$

so that the outgoing component of intensity is given by

$$\mathbf{I}\cdot\mathbf{n} = \frac{c}{2\pi}\mathscr{H}\cdot\mathscr{H}^* = \frac{ck^4}{2\pi r^2}|\mathbf{n}\times\boldsymbol{\mu}(\omega)|^2. \tag{2.9.23}$$

Denoting by θ the polar angle with respect to $\boldsymbol{\mu}(\omega)$ and using (2.9.8) we can write

$$\mathbf{I}\cdot\mathbf{n} = I(n,\theta) = \frac{\omega^4}{2\pi c^3 r^2}|\boldsymbol{\mu}(\omega)|^2\sin^2\theta \tag{2.9.24}$$

(absolute value in *both* vector and complex number senses). This represents the well-known intensity profile for dipole radiation.

The total radiated power can be obtained by integrating the intensity

† It is, in fact, a well-known trick that the time average of a product of $F = \mathscr{F}\,\mathrm{e}^{-i\omega t} + \mathscr{F}^*\,\mathrm{e}^{i\omega t}$ with $G = \mathscr{G}\,\mathrm{e}^{-i\omega t} + \mathscr{G}^*\,\mathrm{e}^{i\omega t}$ is given by $\overline{FG} = 2\,\mathrm{Re}\,\mathscr{F}\mathscr{G}^*$

over a closed surface sufficiently distant from the source:

$$P \equiv \int \mathbf{I} \cdot d\boldsymbol{\sigma} = \int_0^{2\pi} \int_0^{\pi} I(r, \theta) r^2 \sin\theta \, d\theta \, d\psi. \tag{2.9.25}$$

The result is†

$$P = \frac{4\omega^4}{3c^2} |\boldsymbol{\mu}(\omega)|^2. \tag{2.9.26}$$

By appropriate *ad hoc* adaptation, this classical result can be made applicable to spontaneous emission by an excited quantum system (cf. Section 8.8).

2.10 Plane Electromagnetic Waves

At large distances from a localized radiation source, small segments of the outgoing spherical wavefronts approach planarity. The properties of *plane electromagnetic waves* can thereby be deduced as a limiting case of the preceding results. It is instructive, however, to develop the theory of plane waves independently.

In the absence of sources, where $\rho = 0$ and $\mathbf{j} = 0$, the electromagnetic potentials satisfy homogeneous wave equations. It is convenient to adopt Coulomb gauge‡ wherein the scalar potential is a solution of Laplace's equation (cf. 2.3.6)

$$\nabla^2 \Phi = 0. \tag{2.10.1}$$

Φ accordingly contains no essential time dependence. It represents merely some background to the radiation field and can, with no loss of generality,

† This is equivalent to Larmor's famous formula for the radiation from an accelerated electron:

$$P = 2\, e^2 a^2 / 3c^3.$$

When P represents the time-averaged power then $\mathbf{a}$ is to be interpreted as the rms acceleration of the electron. For harmonic oscillation, $\mathbf{a} = -\omega^2 \mathbf{r}$. The dipole moment of the electron is given by $\boldsymbol{\mu} = -e\,\mathbf{r}$. Now

$$\boldsymbol{\mu} = \boldsymbol{\mu}(\omega)\, e^{-i\omega t} + \boldsymbol{\mu}^*(\omega)\, e^{i\omega t}$$

so that $\overline{\mu^2} = 2|\boldsymbol{\mu}(\omega)|^2$. Thus $|\boldsymbol{\mu}(\omega)|^2 = e^2 \overline{a^2}/2\omega^4$, which converts (2.9.26) into Larmor's expression.

‡ In Lorentz gauge, the homogeneous wave equations corresponding to (2.3.11) and (2.3.12) read

$$\Box \Phi = 0, \qquad \Box \mathbf{A} = 0$$

their solutions being coupled by the Lorentz condition (2.3.9). If, in some Lorentz frame, the radiation source represents a steady-state distribution:

$$\rho = 0, \qquad \nabla \cdot \mathbf{j} = 0,$$

the Lorentz gauge potentials are identical with (2.10.2).

be normalized to zero. The electromagnetic potentials hence conform to the conditions

$$\Phi = 0, \qquad \nabla \cdot \mathbf{A} = 0. \tag{2.10.2}$$

The vector potential in free space becomes thereby (cf. 2.3.3) a divergenceless solution of the wave equation:

$$\Box \mathbf{A} = 0, \qquad \nabla \cdot \mathbf{A} = 0. \tag{2.10.3}$$

The associated electric and magnetic radiation fields† are found from

$$\mathbf{E} = -\frac{1}{c}\frac{\partial \mathbf{A}}{\partial t} \tag{2.10.4}$$

and

$$\mathbf{H} = \nabla \times \mathbf{A}. \tag{2.10.5}$$

The simplest wavelike solutions to (2.10.3) are of the form

$$\mathbf{A}(\mathbf{r}, t) = \mathscr{A}_\omega \exp\left(i(\mathbf{k}\cdot\mathbf{r} - \omega t)\right) + \text{cc} \tag{2.10.6}$$

with

$$\mathscr{A}_\omega = |\mathscr{A}_\omega| e^{i\alpha}, \tag{2.10.7}$$

corresponding to monochromatic plane waves propagating in the direction of the *wavevector* **k**. The magnitude of **k** has the same significance as before, namely,

$$k = |\mathbf{k}| = \omega/c \tag{2.10.8}$$

The Coulomb gauge condition applied to (2.10.6) gives

$$\nabla \cdot \mathbf{A}(\mathbf{r}, t) = i\mathbf{k}\cdot\mathscr{A}_\omega \exp\left(i(\mathbf{k}\cdot\mathbf{r} - \omega t)\right) = 0 + \text{cc} \tag{2.10.9}$$

This implies

$$\mathbf{k}\cdot\mathbf{A} = 0, \tag{2.10.10}$$

showing that oscillations in **A** are *transverse* to the direction of propagation.‡ When $\mathscr{A}_\omega$ is, in addition, constant in direction, as in (2.10.7), the wave is characterized as linearly polarized.§

† The fields are themselves divergenceless solutions of the wave equation, viz,

$$\Box \mathbf{E} = 0, \quad \nabla \cdot \mathbf{E} = 0 \quad \text{and} \quad \Box \mathbf{H} = 0, \quad \nabla \cdot \mathbf{H} = 0.$$

‡ Coulomb gauge is, for this reason, also called *transverse gauge*.

§ A more general state of polarization can be represented by

$$\mathscr{A} = \mathbf{e}_1 \mathscr{A}_1 + \mathbf{e}_2 \mathscr{A}_2$$

where $\mathbf{e}_1$ and $\mathbf{e}_2$ are orthogonal unit vectors normal to **k** and

$$\mathscr{A}_1 = |\mathscr{A}_1| \exp(i\alpha_1), \qquad \mathscr{A}_2 = |\mathscr{A}_2| \exp(i\alpha_2).$$

When $|\alpha_1 - \alpha_2| = 0$ or π, the wave is *linearly polarized*, as above. Otherwise, it is *elliptically polarized*. The special case $|\mathscr{A}_1| = |\mathscr{A}_2|$ and $|\alpha_1 - \alpha_2| = \pi/2$ corresponds to *circular polarization*.

The electric and magnetic fields associated with the radiation, since they satisfy analogous wave equations are also transverse oscillations, in agreement with the results of the preceding sections. More specifically, using (2.10.6) in (2.10.4) and (2.10.5), we find

$$\left.\begin{aligned} \mathbf{E} &= ik\mathscr{A}_\omega \exp(i(\mathbf{k}\cdot\mathbf{r} - \omega t)) + \text{cc} \\ &\text{and} \\ \mathbf{H} &= i\mathbf{k} \times \mathscr{A}_\omega \exp(i(\mathbf{k}\cdot\mathbf{r} - \omega t)) + \text{cc}. \end{aligned}\right\} \quad (2.10.11)$$

We arrive thereby at the familiar picture of electromagnetic waves as transverse synchronous oscillations of mutually perpendicular electric and magnetic fields. The electric field vibrates in the plane of polarization, defined by $\mathscr{A}_\omega$. The waves advance at a speed c in the direction of $\mathbf{k}$, perpendicular to both $\mathbf{E}$ and $\mathbf{H}$ in the sense $\mathbf{E} \times \mathbf{H}$. In gaussian units (although not in SI units), the electric and magnetic radiation fields in free space are of equal magnitude:

$$|\mathbf{E}(\mathbf{r}, t)| = |\mathbf{H}(\mathbf{r}, t)|. \quad (2.10.12)$$

The intensity of the plane wave represented by (2.10.11) is given by

$$\mathbf{I} = \frac{c}{4\pi}\overline{\mathbf{E} \times \mathbf{H}} = \frac{c}{2\pi}k\mathscr{A}_\omega \times (\mathbf{k} \times \mathscr{A}_\omega^*) \quad (2.10.13)$$

having used the footnote on p 42 to get the time average. (Averaging over space gives the same result here.) The magnitude of (2.10.13), noting (2.10.8), is

$$I = \frac{\omega^2}{2\pi c}|\mathscr{A}_\omega|^2 \quad (2.10.14)$$

Alternatively, in terms of the actual vector potential (cf. 2.10.6),

$$I = \frac{\omega^2 \overline{A^2}}{4\pi c}. \quad (2.10.15)$$

Every real radiation source produces a finite—however narrow— band of frequencies. We shall therefore require an expression for the intensity of a polychromatic plane wave. The vector potential can be represented by generalization of (2.10.6) to a Fourier integral

$$\mathbf{A}(\mathbf{r}, t) = \int_{-\infty}^{\infty} \mathscr{A}(\omega) \exp(i(\mathbf{k}\cdot\mathbf{r} - \omega t))\, d\omega \quad (2.10.16)$$

in which contributing wavevectors have the same direction but varying magnitudes $k = \omega/c$. Again, for $\mathbf{A}(\mathbf{r}, t)$ to be a real function, the negative-frequency components in (2.10.16) must satisfy

$$\mathscr{A}(-\omega) = \mathscr{A}^*(\omega). \quad (2.10.17)$$

In the particular application to be made in Section 8.5, we shall require (2.10.16) to represent a finite pulse of radiation during the time interval Δt from t_0 to t. The Fourier amplitudes $\mathscr{A}(\omega)$ must accordingly be chosen such that

$$\mathbf{A}(\mathbf{r}, t') = 0 \quad \text{for} \quad t' < t_0 \quad \text{and} \quad t' > t. \tag{2.10.18}$$

The electric and magnetic fields are given by obvious generalization of (2.10.11):

$$\left.\begin{aligned} \mathbf{E} &= \int_{-\infty}^{\infty} \mathrm{i}k\,\mathscr{A}(\omega)\exp\left(i(\mathbf{k}\cdot\mathbf{r} - \omega t)\right)\mathrm{d}\omega \\ \mathbf{H} &= \int_{-\infty}^{\infty} i\mathbf{k} \times \mathscr{A}(\omega)\exp\left(i(\mathbf{k}\cdot\mathbf{r} - \omega t)\right)\mathrm{d}\omega \end{aligned}\right\} \tag{2.10.19}$$

The energy flux is thus the double integral

$$\mathbf{S} = \frac{c}{4\pi}\mathbf{E} \times \mathbf{H} = \frac{c}{4\pi}\mathbf{E} \times \mathbf{H}^* = \frac{c}{4\pi}\int_{-\infty}^{\infty}\int_{-\infty}^{\infty} k\mathscr{A}(\omega)$$
$$\times \left[\mathbf{k}' \times \mathscr{A}^*(\omega')\right]\exp\left(i(\mathbf{k} - \mathbf{k}')\cdot\mathbf{r}\right)\exp\left(i(\omega' - \omega)t\right)\mathrm{d}\omega\,\mathrm{d}\omega'. \tag{2.10.20}$$

Now the appropriate intensity is the average over the time interval Δt:

$$\mathbf{I} = \frac{1}{\Delta t}\int_{t_0}^{t}\mathbf{S}(t')\,\mathrm{d}t' = \frac{1}{\Delta t}\int_{-\infty}^{\infty} S(t')\,\mathrm{d}t' \tag{2.10.21}$$

the limits of integration being extended to $\pm\infty$ by virtue of (2.10.19). Putting (2.10.20) into (2.10.21) gives the deltafunction (cf. A.15):

$$\int_{-\infty}^{\infty} \mathrm{e}^{i(\omega' - \omega)t'}\,\mathrm{d}t' = 2\pi\delta(\omega' - \omega).$$

Thus†

$$\mathbf{I} = \frac{c}{2\Delta t}\int_{-\infty}^{\infty} k\mathscr{A}(\omega)[\mathbf{k} \times \mathscr{A}^*(\omega)]\,\mathrm{d}\omega. \tag{2.10.22}$$

In magnitude,

$$I = \frac{1}{2c\Delta t}\int_{-\infty}^{\infty}\omega^2|\mathscr{A}(\omega)|^2\,\mathrm{d}\omega = \frac{1}{c\Delta t}\int_{0}^{\infty}\omega^2|\mathscr{A}(\omega)|^2\,\mathrm{d}\omega \tag{2.10.23}$$

† This could also have been obtained by applying the appropriate form of Parceval's theorem (cf. B.15):

$$\int_{-\infty}^{\infty}\mathbf{E}(t) \times \mathbf{H}^*(t)\,\mathrm{d}t = 2\pi\int_{-\infty}^{\infty}\mathscr{E}(\omega) \times \mathscr{H}^*(\omega)\,\mathrm{d}\omega$$

in which $\mathscr{E}(\omega)$ and $\mathscr{H}(\omega)$ are the Fourier amplitudes in (2.10.19).

the negative frequencies having been eliminated by using (2.10.17). The total intensity thus has the form of a superposition of individual frequency components:

$$I = \int_0^\infty I(\omega)\,\mathrm{d}\omega \tag{2.10.24}$$

with

$$I(\omega) = \frac{\omega^2}{c\Delta t}|\mathscr{A}(\omega)|^2. \tag{2.10.25}$$

Note that $I(\omega)$ represents the average radiation energy transferred across unit area in unit time *per unit frequency interval.*

The intensity can also be expressed in terms of the electric or the magnetic field as follows:

$$I(\omega) = \frac{c}{\Delta t}|\mathscr{E}(\omega)|^2 = \frac{c}{\Delta t}|\mathscr{H}(\omega)|^2 \tag{2.10.26}$$

in which $\mathscr{E}(\omega)$ and $\mathscr{H}(\omega)$ are the Fourier amplitudes in (2.10.18). The latter two vectors are, of course, perpendicular but equal in magnitude. Equations (2.10.25) and (2.10.26) will find application in our later discussion of radiative transitions.

3

Survey of Quantum Mechanical Formalism

3.1 Approaches to Quantum Mechanics

Modern quantum theory began with two independent mathematical formulations: Heisenberg's *matrix mechanics* (1925)† and Schrödinger's *wave mechanics* (1926).‡ Dirac's *transformation theory*§ provided a unified synthesis, encompassing both wave and matrix mechanics as specific representations. More recently, elegant reformulations of quantum mechanics have been inspired by developments in quantum electrodynamics and field theory. Best known among these are Feynman's path-integral approach|| and Schwinger's action principle.¶.

Schrödinger's wave mechanics has proven to be the most appropriate basis for the theory of atoms and molecules. Concepts and notation drawn from matrix mechanics and transformation theory nevertheless find extensive utilization in molecular physics and quantum chemistry. In the following sections, wave-mechanical theory will be formulated in terms of a set of five fundamental postulates, designed to provide a convenient framework for our study of nonrelativistic time-dependent phenomena. As regards the postulational system itself, however, neither uniqueness, absolute rigor nor logical elegance can be claimed.

3.2 Wavefunctions and Integrals

Postulate 1. The state of a quantum system is specified by a wavefunction $\Psi(q_1 \ldots q_n, t)$. Here $q_1 \ldots q_n$ are generalized coordinates representing n degrees of freedom (including, if necessary, spins) and t is the time. For brevity, we shall often write $\Psi(q, t)$, with q symbolizing the entire set of

† W. Heisenberg, *Z. Phys.* **33**, 879 (1925); M. Born and P. Jordan, *Z. Phys.* **34**, 858 (1925); M. Born, W. Heisenberg and P. Jordan, *Z. Phys.* **35**, 557 (1925).
‡ E. Schrödinger, *Ann. Phys.* **79**, 361, 489 (1926); **80**, 437 (1926); **81**, 109 (1926).
§ P. A. M. Dirac, "Quantum Mechanics" 4th Ed., Oxford University Press, 1958.
|| R. P. Feynman, *Rev. Mod. Phys.* **20**, 367 (1948).
¶ J. Schwinger, *Phys. Rev.* **82**, 914 (1951); **91**, 713 (1953).

configuration variables. The wavefunction is required to be single-valued, finite† and continuous (in brief, "well-behaved") for all values of the generalized coordinates and time.

A system is said to be in a *stationary state* when its observable properties do not change with time. Wavefunctions representing stationary states have time-dependence which can be factorized out (cf. Section 4.10B). The remaining configuration-dependent function—which we shall designate $\psi(q)$ or $\phi(q)$—then fully describes the system. We shall be concerned in what follows with both time-dependent and time-independent wavefunctions. We may, upon occasion, omit explicit reference to the time dependence when it is of secondary importance.

There are two complementary points of view as to the physical significance of the wavefunction. Schrödinger, preferring the wave-like picture of matter, regarded $\Psi(q, t)$ simply as the amplitude of matter waves. In common with the theories of classical wave phenomena, the square of this amplitude represents a measure of intensity. Since quantum mechanics admits of complex wavefunctions, the obvious generalization for matter-wave intensity is the modulus squared, that is,

$$\rho(q, t) = |\Psi(q, t)|^2 = \Psi^*(q, t)\,\Psi(q, t). \tag{3.2.1}$$

Born, maintaining the particulate (corpuscular) view of matter, interpreted $\rho(q, t)$ as a statistical quantity: the *probability density* for instantaneous occurrence of configuration q at time t. Since the integrated probability of finding the system in *some* configuration is unity, the following normalization condition is indicated‡:

$$\int \Psi^*(q, t)\,\Psi(q, t)\,d\tau = 1. \tag{3.2.2}$$

Convergence of the normalization integral requires a quantum state largely localized within some finite region of configuration space. There occur, as well, states for which the normalization integral diverges. These are generally characterized by infinite delocalization of the system—for example, momentum eigenstates of the free particle (cf. Section 5.1). For such wavefunctions, normalization in terms of the deltafunction can be utilized (cf. Section 3.8).

The notation $\int \ldots d\tau$ will be used throughout to represent integration

† Exceptions to the finiteness condition occur in the relativistic treatment of the hydrogen atom, solutions to the Dirac equation for $j = \frac{1}{2}$ exhibiting weak singularities at the origin.

‡ If the wavefunction is not normalized, the probability density can be represented by

$$\rho(q, t) = \frac{\Psi^*(q, t)\Psi(q, t)}{\int \Psi^*(q, t)\Psi(q, t)\,d\tau}. \tag{3.2.1'}$$

over the set of configuration variables $q_1 \ldots q_n$ (but *not* t), with appropriate metric coefficients. For the space coordinates of a single particle, for example, $d\tau = dx\,dy\,dz$ in cartesian coordinates.† In generalized coordinates

$$d\tau = g(q_1, q_2, q_3)\, dq_1\, dq_2\, dq_3 \tag{3.2.3}$$

where g is the Jacobian of the transformation from x, y, z to q_1, q_2, q_3:

$$g(q_1, q_2, q_3) \equiv \begin{vmatrix} \partial x/\partial q_1 & \partial x/\partial q_2 & \partial x/\partial q_3 \\ \partial y/\partial q_1 & \partial y/\partial q_2 & \partial y/\partial q_3 \\ \partial z/\partial q_1 & \partial z/\partial q_2 & \partial z/\partial q_3 \end{vmatrix} \tag{3.2.4}$$

In spherical polar coordinates, for example, $(q_1 = r, q_2 = \theta,\ q_3 = \phi)$, $g = r^2 \sin\theta$ and $d\tau = r^2 \sin\theta\, dr\, d\theta\, d\phi$.

For the general case of n degrees of freedom, the element of integration can be denoted.

$$d\tau = g(q)\, dq, \ldots dq_n, \tag{3.2.5}$$

$g(q)$ representing the Jacobian connecting the n generalized coordinates to an appropriate set of n cartesian variables.

It is convenient, for subsequent considerations, to introduce the following abbreviated notation for integrals:

$$(f, g) \equiv \int f^* g\, d\tau. \tag{3.2.6}$$

This notation—called the *inner product* of f and g— is suggested by the representation of functions as vectors in Hilbert space (cf. Section 3.10). Consistent with the definition (3.2.6),

$$(g, f) = \int g^* f\, d\tau = (f, g)^*. \tag{3.2.7}$$

The normalization condition (3.2.2) can accordingly be written

$$(\Psi, \Psi) = 1. \tag{3.2.8}$$

The *norm* of a function is defined by

$$\|f\| \equiv (f, f)^{\frac{1}{2}}, \tag{3.2.9}$$

a function being thus *normalized* if $\|f\| = 1$. In analogy with the case of perpendicular vectors, two functions are said to be *orthogonal* if

$$(f, g) = 0. \tag{3.2.10}$$

3.3 Operators

Postulate 2. Every dynamical variable for a quantum system is represented by

† We alternatively use $d^3\mathbf{r}$ to represent a 3-dimensional volume element.

a linear hermitian operator. An operator represents a prescription for transforming one function into another.† Symbolically this can be written

$$\mathscr{A}\Psi(q, t) = \Phi(q, t). \tag{3.3.1}$$

From a physical point of view, the action of an operator on a wavefunction Ψ can be associated with the process of measuring the corresponding observable. The transformed wavefunction, say Φ in eqn (3.3.1), can correspondingly be pictured as the state of the system *after* the measurement. That Φ is, in general, different from Ψ is consistent with the fact that the process of measurement generally produces an irreducible perturbation of a quantum system.

An operator $\mathscr{A}$ is *linear* if it obeys the distributive law

$$\mathscr{A}(c_1 f + c_2 g) = c_1 \mathscr{A} f + c_2 \mathscr{A} g, \tag{3.3.2}$$

for arbitrary functions f and g and arbitrary constants c_1 and c_2. For example, $\mathscr{A} = \partial/\partial x$ is a linear operator while $\mathscr{A} = \log$ is not.

Integrals of the general form $\int f^* \mathscr{A} g \, \mathrm{d}\tau$ are of fundamental importance in quantum theory. In terms of the notation of eqn (3.2.6), these integrals can be represented using

$$(f, \mathscr{A}g) = \int f^* \mathscr{A} g \, \mathrm{d}\tau. \tag{3.3.3}$$

The *adjoint* $\mathscr{A}^\dagger$ of an operator $\mathscr{A}$ is defined by

$$(f, \mathscr{A}g) = (g, \mathscr{A}^\dagger f)^*. \tag{3.3.4}$$

An operator is *hermitian* or *self-adjoint* with respect to a particular set or class of functions if

$$(f, \mathscr{A}g) = (\mathscr{A}f, g) = (g, \mathscr{A}f)^*, \tag{3.3.5}$$

for arbitrary functions f and g in that set. Symbolically, then

$$\mathscr{A}^\dagger = \mathscr{A}. \tag{3.3.6}$$

The manipulation implied by (3.3.5), in which the action of a hermitian operator is turned from one function in the integrand to the other, is sometimes known as the "turnover rule".

A hermitian operator can be constructed from an arbitrary non-hermitian operator by linear combination with its adjoint. Thus $\frac{1}{2}(\mathscr{A} + \mathscr{A}^\dagger)$ and $(i/2)(\mathscr{A} - \mathscr{A}^\dagger)$ are hermitian.

An alternative way of representing a linear operator makes use of an integral transform. Instead of (3.3.1), the fundamental defining relationship is

$$\mathscr{A}\psi(q) = \int A(q, q')\psi(q') \, \mathrm{d}\tau'. \tag{3.3.7}$$

† Analogously, a function represents a prescription for tuning one *number* into another.

Each operator $\mathscr{A}$ is thereby uniquely associated with a *kernel*, $A(q, q')$, a function twice over of the configuration variables. By setting $\psi(q) = \delta(q - q')$ we also find the formal relation

$$\mathscr{A}\delta(q - q') = A(q, q'). \tag{3.3.8}$$

For a hermitian operator, the kernel fulfils the condition

$$A(q, q') = A(q', q)^* \tag{3.3.9}$$

It is interesting to note that the deltafunction plays the role of kernel for the identity operator, for, by (A.33),

$$\psi(q) = \int \delta(q - q')\psi(q')\,\mathrm{d}\tau'. \tag{3.3.10}$$

The differential operator $\partial/\partial q_i$ can be represented by the kernel $-\delta'(q_i - q_i')$, since, from (A.27) and noting that $\mathrm{d}\tau'$ contains the factor $\mathrm{d}q_i'$,

$$\partial\psi(q)/\partial q_i = -\int \delta'(q_i - q_i')\psi(q')\,\mathrm{d}\tau'. \tag{3.3.11}$$

In general, by virtue of (A.28),

$$\partial^n\psi(q)/\partial q_i^n = (-1)^n \int \delta^{(n)}(q_i - q_i')\psi(q')\,\mathrm{d}\tau'. \tag{3.3.12}$$

The product of two operators, say $\mathscr{A}\mathscr{B}$, represents the successive action of the operators reading *right* to *left*—i.e., first $\mathscr{B}$, then $\mathscr{A}$. The operator $\mathscr{A}^N$ represents the result of N successive applications of $\mathscr{A}$. In general, operators do not commute, meaning that $\mathscr{A}\mathscr{B} \neq \mathscr{B}\mathscr{A}$. In terms of the discussion following eqn (3.3.1), the noncommutativity of operators can be associated with the perturbing effect one measurement on a quantum system might have on subsequent measurements.

The commutator of two operators is defined by

$$[\mathscr{A}, \mathscr{B}] \equiv \mathscr{A}\mathscr{B} - \mathscr{B}\mathscr{A}. \tag{3.3.13}$$

When $[\mathscr{A}, \mathscr{B}] = 0$, the two operators are said to *commute*. The following identities among commutators are easily demonstrated:

$$[\mathscr{A}, \mathscr{A}] = 0 \tag{3.3.14}$$

$$[\mathscr{A}, \mathscr{B}] = -[\mathscr{B}, \mathscr{A}] \tag{3.3.15}$$

$$[\mathscr{A}, \mathscr{B} + \mathscr{C}] = [\mathscr{A}, \mathscr{B}] + [\mathscr{A}, \mathscr{C}] \tag{3.3.16}$$

$$[\mathscr{A}^2, \mathscr{B}] = \mathscr{A}[\mathscr{A}, \mathscr{B}] + [\mathscr{A}, \mathscr{B}]\mathscr{A} \tag{3.3.17}$$

$$[\mathscr{A}, \mathscr{B}\mathscr{C}] = [\mathscr{A}, \mathscr{B}]\mathscr{C} + \mathscr{B}[\mathscr{A}, \mathscr{C}] \tag{3.3.18}$$

$$[[\mathscr{A}, \mathscr{B}], \mathscr{C}] + [[\mathscr{B}, \mathscr{C}], \mathscr{A}] + [[\mathscr{C}, \mathscr{A}], \mathscr{B}] = 0 \tag{3.3.19}$$

If $\mathscr{A}$ and $\mathscr{B}$ are each hermitian operators, their product will be hermitian only if $\mathscr{A}$ and $\mathscr{B}$ commute. However, the combinations $\mathscr{A}\mathscr{B} + \mathscr{B}\mathscr{A}$ and $i[\mathscr{A}, \mathscr{B}]$ will invariably be hermitian.

3.4 Quantum Conditions

The forms of the operators representing quantum-mechanical observables are most conveniently deduced from the corresponding classical dynamical variables. A very general postulational principle relates Poisson brackets to operator commutators. The *Poisson bracket* of two dynamical variables is defined by (cf. 1.3.3):

$$\{A, B\} \equiv \sum_i \left(\frac{\partial A}{\partial q_i}\frac{\partial B}{\partial p_i} - \frac{\partial B}{\partial q_i}\frac{\partial A}{\partial p_i}\right), \tag{3.4.1}$$

the summation running over all degrees of freedom of the dynamical system. *Postulate 3. Every pair of quantum-mechanical operators satisfies the commutation relation implied by the correspondence*

$$i\hbar\{A, B\} \to [\mathscr{A}, \mathscr{B}] \tag{3.4.2}$$

This relationship is made plausible by the correspondence between the algebraic properties of Poisson brackets and commutators: compare (3.3.14)–(3.3.19) with (1.3.4)–(1.3.9) (with appropriate modification of (1.3.7) and (1.3.8) for the order of factors).

By choosing A to be one of the generalized coordinates and B, one of the generalized momenta, we have

$$\{q_i, p_j\} = \delta_{ij}. \tag{3.4.3}$$

This implies, for the corresponding operators,

$$[q_i, p_j] = i\hbar\, \delta_{ij}. \tag{3.4.4}$$

Also, it is easily shown that

$$[q_i, q_j] = 0, \qquad [p_i, p_j] = 0, \qquad \text{all } i, j. \tag{3.4.5}$$

Relations (3.4.4) and (3.4.5) are called the *fundamental quantum conditions*, these showing where the noncommutativity among quantum-mechanical operators intrinsically lies.

The linear operators $-i\hbar\partial/\partial q_i$ satisfy the same commutation relations with respect to the q_i as do the p_i, namely (3.4.4) and (3.4.5). The quantization prescription $q_i \to q_i$, $p_i \to -i\hbar\partial/\partial q_i$ to construct generalized coordinate and momentum operators is thereby strongly suggested.† Unless the q_i are

† This is the basis of *Schrödinger's representation*. Another possible choice, although much less often used, is the *momentum representation*, based on $p_i \to p_i$, $q_i \to i\hbar\partial/\partial p_i$.

cartesian coordinates, the momentum operators, as defined, might not be hermitian. Hermiticity can, however, be ensured by the following procedure.† Clearly, the operators

$$-i\hbar[\partial/\partial q_i + f_i(q_i)], \tag{3.4.6}$$

where $f_i(q_i)$ is an arbitrary function of q_i, will also satisfy the same commutation relations as the p_i. If these operators are to be hermitian, they must satisfy [cf. (3.3.13), using (3.2.5) for $d\tau$]

$$\int \psi^*(q)\{-i\hbar[\partial/\partial q_i + f_i(q_i)]\,\phi(q)\}\,g(q)\,dq_1 \ldots dq_n$$
$$= \int \{-i\hbar[\partial/\partial q_i + f_i(q_i)]\,\psi(q)\}^*\,\phi(q)\,g(q)\,dq_1 \ldots dq_n, \tag{3.4.7}$$

$\psi(q)$ and $\phi(q)$ being arbitrary wavefunctions depending on $q_1 \ldots q_n$. Integrating by parts on either side of the equation and assuming that all boundary contributions vanish,‡ it is found that (3.4.7) is satisfied provided that

$$f_i(q_i) = \frac{1}{2g}\frac{\partial g}{\partial q_i}\,. \tag{3.4.8}$$

(This is indeed a function of q_i alone since, for any system of orthogonal generalized coordinates, all the $q_1 \ldots q_n$ other than q_i will occur as identical factors in g and $\partial g/\partial q_i$.) Incorporating (3.4.8) into (3.4.6), the operator can be more compactly represented $-i\hbar g^{-\frac{1}{2}}\partial/\partial q_i g^{\frac{1}{2}}$. We arrive thereby at the canonical quantization prescription for each pair of conjugate generalized coordinates and momenta, viz.,

$$q_i \to q_i, \qquad p_i \to -i\hbar g^{-\frac{1}{2}}\,\partial/\partial q_i\,g^{\frac{1}{2}}, \tag{3.4.9}$$

the coordinate variables becoming formally multiplicative operators. This prescription is simplest, of course, in cartesian coordinates, the momentum components being given by

$$p_x \to -i\hbar\,\partial/\partial x, \qquad p_y \to -i\hbar\,\partial/\partial y, \qquad p_z \to -i\hbar\,\partial/\partial z. \tag{3.4.10}$$

In vector form,

$$\mathbf{p} \to -i\hbar\nabla. \tag{3.4.11}$$

The momentum squared is thus

$$p^2 \to -\hbar^2\nabla^2, \tag{3.4.12}$$

a result valid, in fact, for any coordinate system.

† For further discussion, see B. Podolsky, *Phys. Rev.* **32**, 812 (1928); P. D. Robinson and J. O. Hirschfelder, *J. Math. Phys.*, **4**, 338 (1963).

‡ The hermiticity of the operator is therefore defined with respect to the class of functions, including $\psi(q)$ and $\phi(q)$, for which these boundary terms vanish.

It can thus be generalized, for a dynamical variable A having a classical analog, the corresponding quantum-mechanical operator $\mathscr{A}$ can be constructued by first expressing A as a function of $q_1 \ldots q_n, p_1 \ldots p_n$, and (if necessary) t and then applying (3.4.9). The general quantization prescription in Schrödinger's representation can thus be symbolized:

$$A(q_1 \ldots q_n, p_1 \ldots p_n, t) \rightarrow$$

$$\mathscr{A}(q_1 \ldots q_n, -i\hbar g^{-\frac{1}{2}}\, \partial/\partial q_1 g^{\frac{1}{2}}, \ldots -i\hbar g^{-\frac{1}{2}} \partial/\partial q_n g^{\frac{1}{2}}, t) \tag{3.4.13}$$

Should any ambiguity arise as to the order of non-commuting factors in $\mathscr{A}$, a linear combination should be constructed such that the resulting operator is hermitian.† For example, the quantum analog of the classical quantity p^2q can be one of the following:

$$\tfrac{1}{2}(p^2q + qp^2), \qquad \tfrac{1}{2}i(p^2q - qp^2), \qquad pqp,$$

or some linear combination thereof.

The operators most commonly dealt with do not explicitly depend on time, an important exception being, however, the Hamiltonian representing interaction of matter with a radiation field.

Quantization of a set of dynamical variables is fundamentally a matter to be considered individually for each dynamical system. The method of classical analogy described above will not suffice, of course, when observables lack classical analogues (e.g., spin). In such instances, however, analogies with known operators (e.g., orbital angular momenta) will generally provide sufficient guidelines.

3.5 The Hamiltonian

This dynamical variable is of central significance in both classical and quantum mechanics. In most cases of interest, the Hamiltonian is equal to the energy expressed as a function of generalized coordinates and momenta. For example, the classical Hamiltonian function for a particle in a conservative potential field, viz.,

$$H(\mathbf{r}, \mathbf{p}) = \frac{p^2}{2m} + V(\mathbf{r}), \tag{3.5.1}$$

is transformed using (3.4.12) to the Hamiltonian operator

$$\mathscr{H} = -\frac{\hbar^2}{2m}\nabla^2 + V(\mathbf{r}). \tag{3.5.2}$$

When non-cartesian coordinates are involved, construction of the Hamiltonian can be a little more complicated. As an illustration, consider the

† H. Weyl, *Z. Phys.* **46**, 1 (1927); R. McCoy, *Proc. Nat. Acad. Sci.* **18**, 674 (1932).

motion of a particle in a central field, expressed in spherical polar coordinates. The classical Lagrangian function $L = T - V$, transformed into spherical coordinates, has the form

$$L = \frac{m}{2}(\dot{r}^2 + r^2\dot{\theta}^2 + r^2 \sin^2 \theta\dot{\phi}^2) - V(r). \tag{3.5.3}$$

The generalized momenta are accordingly (cf. 1.2.8)

$$\begin{aligned} p_r &= \partial L/\partial \dot{r} = m\dot{r} \\ p_\theta &= \partial L/\partial \dot{\theta} = mr^2\dot{\theta} \\ p_\phi &= \partial L/\partial \dot{\phi} = mr^2 \sin^2 \theta\phi, \end{aligned} \tag{3.5.4}$$

which leads, using (1.2.10), to the Hamiltonian function

$$H = \frac{1}{2m}\left(p_r^2 + \frac{p_\theta^2}{r^2} + \frac{p_\phi^2}{r^2 \sin^2 \theta}\right) + V(r). \tag{3.5.5}$$

To quantize (3.5.5), make use of (3.4.9) with $g = r^2 \sin\theta$ to get the momentum operators:

$$\begin{aligned} p_r &= -i\hbar\left(\frac{\partial}{\partial r} + \frac{1}{r}\right) \\ p_\theta &= -i\hbar\left(\frac{\partial}{\partial \theta} + \tfrac{1}{2}\operatorname{cotn}\theta\right) \\ p_\phi &= -ih\frac{\partial}{\partial \phi} \end{aligned} \tag{3.5.6}$$

Thus, after some rearranging of differential operators,

$$\mathscr{H} = -\frac{\hbar^2}{2m}\left(\frac{1}{r^2}\frac{\partial}{\partial r}r^2\frac{\partial}{\partial r} + \frac{1}{r^2 \sin\theta}\frac{\partial}{\partial \theta}\sin\theta\frac{\partial}{\partial \theta} + \frac{1}{r^2 \sin^2 \theta}\frac{\partial^2}{\partial \phi^1}\right) + V(r), \tag{3.5.7}$$

which, of course, corresponds to (3.5.2) expressed in spherical coordinates.

3.6 Expectation Values of Dynamical Variables

Measurement of some dynamical variable $\mathscr{A}$ on a quantum system in a state $\Psi(q, t)$ will, in general, yield nonreproducible individual results. However, in a sufficiently large number of measurements carried out on identically-prepared systems, a definite and reproducible *statistical* pattern will emerge. This statistical behaviour can be characterized by a probability distribution function $\rho(a)$, giving the relative frequency of observing each possible result.

If the possible values of a comprise a continuum, with $\rho(a)$ normalized according to

$$\int \rho(a)\, da = 1, \tag{3.6.1}$$

the *average* or *expectation value* of a series of measurements is given by

$$\langle \mathscr{A} \rangle \text{ or } \bar{a} = \int a\rho(a)\, da. \tag{3.6.2}$$

The case of a discrete distribution is formally contained, as well, for, if the value a_n occurs with probability p_n, the distribution function can be represented

$$\rho(a) = \sum_n p_n \delta(a - a_n). \tag{3.6.3}$$

This gives, for (3.6.1),

$$\sum_n p_n = 1 \tag{3.6.4}$$

and, for (3.6.2),

$$\bar{a} = \sum_n a_n p_n \,. \tag{3.6.5}$$

Further characterization of the distribution is contained in its *moments*—expectation values of powers of $\mathscr{A}$, viz,

$$\langle \mathscr{A}^{\mathrm{N}} \rangle \quad \text{or} \quad \overline{a^{\mathrm{N}}} = \int a^{\mathrm{N}} \rho(a)\, da, \tag{3.6.6}$$

with $\mathrm{N} = 2, 3 \ldots$ According to a well-known principle in statistics, knowledge of all the moments suffices, under some very general conditions, to uniquely determine a distribution. By generalization of (3.6.6), the expectation value of $f(a)$, any analytic function of a, is given by

$$\langle f(a) \rangle = \int f(a)\rho(a)\, da. \tag{3.6.7}$$

Apart from the average value $\bar{a}$, the most significant property of a distribution is the standard deviation, Δa, defined as follows:

$$(\Delta a)^2 \equiv \int (a - \bar{a})^2 \rho(a)\, da = \overline{a^2} - \bar{a}^2. \tag{3.6.8}$$

This parameter characterizes the extent to which individual observations are spread about the average value.

Having discussed the general description of distributions and expectation values, we now state *Postulate 4*: *The expectation value of a dynamical* *variable $\mathscr{A}$ in a quantum state $\Psi(q, t)$ is given by*

$$\langle \mathscr{A} \rangle = \frac{\int \Psi^* \mathscr{A} \Psi \, d\tau}{\int \Psi^* \Psi \, d\tau} = \frac{(\Psi, \mathscr{A}\Psi)}{(\Psi, \Psi)}. \tag{3.6.9}$$

When $\mathscr{A}$ and/or Ψ depend on t, then $\langle \mathscr{A} \rangle$ will, in general, be time dependent. When $\Psi(q, t)$ is normalized, the denominator in (3.6.9) equals unity and can be omitted.

The interpretation of the wavefunction as a configuration probability amplitude can be independently surmised on the basis of the expectation-value postulate. Thus, setting $\mathscr{A} = \delta(q - q')$ in (3.6.9), we obtain

$$\langle \delta(q - q') \rangle = \frac{\Psi^*(q', t)\Psi(q', t)}{(\Psi, \Psi)} \tag{3.6.10}$$

which has the significance of $\rho(q', t)$, in agreement with (3.2.1) or (3.2.1′).

When $\mathscr{A}$ has the form of a multiplicative operator—i.e., multiplication by an ordinary function, say $f(q, t) \times$—then (3.6.9) reduces to the classical definition of average value (cf. 3.6.7)

$$\langle f \rangle = \int f(q, t)\rho(q, t) \, d\tau, \tag{3.6.11}$$

having introduced the configuration probability density $\rho(q, t)$. The quantum generalization of (3.6.11) entails replacement of its integrand by the analogous "sandwich" construction $\Psi^* \mathscr{A} \Psi$.

A. *Uncertainty Relations*

The standard deviation for an observable $\mathscr{A}$ can be expressed in quantum-mechanical form as follows (cf. 3.6.8):

$$(\Delta a)^2 = (\Psi, [\mathscr{A} - \bar{a}]^2 \Psi)/(\Psi, \Psi). \tag{3.6.12}$$

The root-mean-square quantity Δa can be pictured as a measure of *uncertainty* associated with a series of observations of $\mathscr{A}$ on a quantum state Ψ. A fundamental quantum-mechanical principle involves the *uncertainty product* $\Delta a \Delta b$ associated with independent observations of two dynamical variable $\mathscr{A}$ and $\mathscr{B}$ on a state Ψ.

A preliminary mathematical result is required. Let u and v be complex functions of the configuration variables q. Then

$$|u - v|^2 = (u^* - v^*)(u - v) \geqslant 0, \tag{3.6.13}$$

for all values of q. Expanding the product gives the equivalent inequality

$$u^*u + v^*v \geqslant u^*v + v^*u. \tag{3.6.14}$$

Supposing now that $u = f/(f, f)^{\frac{1}{2}}$, $v = g/(g, g)^{\frac{1}{2}}$, (3.6.14) becomes

$$\frac{f^*f}{(f, f)} + \frac{g^*g}{(g, g)} \geqslant \frac{f^*g + g^*f}{(f, f)^{\frac{1}{2}}(g, g)^{\frac{1}{2}}}. \tag{3.6.15}$$

Since this is valid for all q, integration over configuration variables will preserve the sense of the inequality. Thereby (noting that (f, f) and (g, g) are just numbers) we obtain

$$(f, f)^{\frac{1}{2}}(g, g)^{\frac{1}{2}} \geqslant \tfrac{1}{2}\int (f^*g + g^*f)\,\mathrm{d}\tau. \tag{3.6.16}$$

This represents, in fact, a generalization of Schwartz' inequality, admitting of complex functions.

With reference to (3.6.16), let us set

$$f = \frac{[\mathscr{A} - \bar{a}]\Psi}{(\Psi, \Psi)^{\frac{1}{2}}}, \qquad g = \frac{i[\mathscr{B} - \bar{b}]\Psi}{(\Psi, \Psi)^{\frac{1}{2}}}. \tag{3.6.17}$$

We have then

$$\begin{aligned}(f, f) &= ([\mathscr{A} - \bar{a}]\Psi, [\mathscr{A} - \bar{a}]\Psi)/(\Psi, \Psi)\\ &= (\Psi, [\mathscr{A} - \bar{a}]^2\Psi)/(\Psi, \Psi) = (\Delta a)^2\end{aligned} \tag{3.6.18}$$

making use of the turnover rule (3.3.5) and the definition (3.6.12). Analogously,

$$(g, g) = (\Delta b)^2 \tag{3.6.19}$$

After some cancellation, it is found that

$$\int (f^*g + g^*f)\,\mathrm{d}\tau = i\int \Psi^*(\mathscr{A}\mathscr{B} - \mathscr{B}\mathscr{A})\Psi\,\mathrm{d}\tau/(\Psi, \Psi) = i\langle[\mathscr{A}, \mathscr{B}]\rangle, \tag{3.6.20}$$

in terms of the expectation value of the commutator.

Substituting (3.6.18), (3.6.19) and (3.6.20) into (3.6.16) yields the *generalized uncertainty relationship*:†

$$\Delta a\,\Delta b \geqslant \frac{i}{2}\langle[\mathscr{A}, \mathscr{B}]\rangle. \tag{3.6.21}$$

If the right-hand side is negative, the result is trivial; to remedy this, interchange $\mathscr{A}$ and $\mathscr{B}$. In any case, it is true that

$$\Delta a\,\Delta b \geqslant \tfrac{1}{2}|\langle[\mathscr{A}, \mathscr{B}]\rangle|. \tag{3.6.22}$$

Better known is the specialization of (3.6.21) to $\mathscr{A} = q_i$, $\mathscr{B} = p_i$. Then, using (3.4.4) with $i = j$, we obtain

$$\Delta\ q_i\,\Delta p_i \geqslant h/4\pi, \tag{3.6.23}$$

† H. P. Robertson, *Phys. Rev.* **35**, 667*A* (1930); E. Schrödinger, *Sitz. preuss. Akad. Wiss.*, p. 296 (1930).

the famous *Heisenberg uncertainty principle*.† Accordingly, a generalized coordinate and its conjugate momentum cannot be simultaneously known exactly. This, in turn, implies that the behaviour of a quantum system must be inherently less deterministically specifiable than a classical trajectory.

3.7 The Eigenvalue Equation

In certain very special circumstances, repeated measurement of a time-independent observable $\mathscr{A}$ on a state Φ_n is found to yield a *reproducible* result, say a_n. Such states are known as *eigenstates* of $\mathscr{A}$. The wavefunctions representing them are called *eigenfunctions*, while the numbers a_n are called *eigenvalues*. The index n is called a *quantum number*; n might also stand for a *set* of quantum numbers. An eigenstate can be characterized by an infinitely sharp distribution function, viz.,

$$\rho(a) = \delta(a - a_n) \tag{3.7.1}$$

so that [cf. (3.6.2), (3.6.6) and (3.6.8)]

$$\bar{a} = a_n, \qquad \overline{a^2} = a_n^2, \qquad \Delta a = 0. \tag{3.7.2}$$

Applying (3.6.12) to the eigenfunction Φ_n, we have

$$(\Delta a)^2 = \frac{(\Phi_n, [\mathscr{A} - a_n]^2 \Phi_n)}{(\Phi_n, \Phi_n)} = 0. \tag{3.7.3}$$

Since $\mathscr{A} - a_n$ is a hermitian operator and $(\Phi_n, \Phi_n) \neq 0$,

$$([\mathscr{A} - a_n]\Phi_n, [\mathscr{A} - a_n]\Phi_n) = \int |(\mathscr{A} - a_n)\Phi_n|^2 \, d\tau = 0. \tag{3.7.4}$$

Now the last integrand is a non-negative function for all values of the coordinates. Vanishing of the integral therefore implies identical vanishing of the integrand, i.e.,

$$(\mathscr{A} - a_n)\Phi_n = 0 \tag{3.7.5}$$

We arrive thereby at the eigenvalue equation

$$\mathscr{A}\Phi_n = a_n \Phi_n. \tag{3.7.6}$$

In many formulations of quantum mechanics, the eigenvalue equation is introduced directly as one of the fundamental postulates. In almost all cases, time dependence can be factored out of the Φ_n, so that the eigenvalue equation also pertains to the time-independent functions $\phi_n(q)$:

$$\mathscr{A}\phi_n = a_n \phi_n \tag{3.7.7}$$

† W. Heisenberg, *Z. Phys.* **43**, 172 (1927).

The acceptable solutions to (3.7.6) or (3.7.7) are generally determined by the analytic restrictions placed on physically-meaningful wavefunctions—that is, that they be single-valued, finite and continuous. When $\mathscr{A}$ is a differential operator, (3.7.6) represents a differential equation and the above restrictions determine boundary conditions on the solutions. Generally, solutions to the eigenvalue equation consistent with its auxiliary conditions are possible only for a restricted range of values of the a_n. The totality of allowed values comprises the *eigenvalue spectrum* of the operator $\mathscr{A}$. It may consist of a finite or infinite discrete set of eigenvalues, a continuum or some combination thereof. For continuum solutions to the eigenvalue equation, we shall use the notation

$$\mathscr{A}\Phi_\nu = a(\nu)\Phi_\nu, \tag{3.7.8}$$

with the quantum number (or numbers) ν running over a continuous range.

The eigenvalues of a dynamical variable represent physically the only possible results of *individual* measurements. Expectation values, being weighted averages of eigenvalues (cf. 3.6.2, 3.6.5), can of course assume arbitrary real values.

Comparing (3.7.6) with (3.3.1), it can be further observed that the eigenfunctions of an operator have the characteristic property of being reproduced —to within a multiplicative constant—by application of the operator. In other words, Ψ in (3.3.1) represents an eigenfunction when $\Phi = \text{const}\ \Psi$. Physically interpreted, this means that measurement of an observable on one of its eigenstates leaves the state of the system unchanged—consistent, in fact, with the presumed reproducibility of that measurement.

The best known eigenvalue equation is, of course, the (time-independent) Schrödinger equation

$$\mathscr{H}\Psi_n = E_n\Psi_n, \tag{3.7.9}$$

the eigenvalues representing the allowed energy levels of the system.

A. *Degeneracy*

Two linearly-independent† eigenfunctions corresponding to the same eigenvalue a_n are said to be *degenerate*. Any d linearly-independent function $\Phi_{n\alpha}$, $\alpha = 1 \ldots d$, satisfying

$$\mathscr{A}\Phi_{n\alpha} = a_n\Phi_{n\alpha} \tag{3.7.10}$$

constitute a d-fold degenerate set of eigenfunctions. Noting that

$$\mathscr{A}\sum_\alpha c_\alpha\Phi_{n\alpha} = \sum_\alpha c_\alpha\mathscr{A}\Phi_{n\alpha} = a_n\sum_\alpha c_\alpha\Phi_{n\alpha}, \tag{3.7.11}$$

† Linear independence is specified since, if $\mathscr{A}$ is a linear operator, any constant multiple of Φ_n is also an eigenfunction for the same eigenvalue. This does not represent degeneracy, however, since const $\times\ \Phi_n$ is not physically distinct from Φ_n.

by virtue of (3.3.2) and (3.7.10), it follows that any linear combination of degenerate eigenfunctions represents itself an eigenfunction with the same eigenvalue. Accordingly, a degenerate set can be constituted in arbitrarily many ways.

B. *Simultaneous Observables*

If two operators $\mathscr{A}$ and $\mathscr{B}$ commute there must exist a set $\{\Phi_n\}$ of simultaneous eigenfunctions of $\mathscr{A}$ and $\mathscr{B}$. This is shown by operating with $\mathscr{B}$ on the eigenvalue equation for $\mathscr{A}$, eqn (3.7.6):

$$\mathscr{B}\mathscr{A}\Phi_n = a_n\mathscr{B}\Phi_n. \tag{3.7.12}$$

But since, by supposition, $\mathscr{B}\mathscr{A} = \mathscr{A}\mathscr{B}$,

$$\mathscr{A}(\mathscr{B}\Phi_n) = a_n(\mathscr{B}\Phi_n). \tag{3.7.13}$$

As emphasized by the inserted brackets, the function $\mathscr{B}\Phi_n$ represents a solution to (3.7.6) corresponding to the eigenvalue a_n. If this is a *nondegenerate* eigenvalue, it must be that $\mathscr{B}\Phi_n$ is just a constant multiple of the known solution Φ_n. Identifying this constant to be the eigenvalue b_b, we can write

$$\mathscr{B}\Phi_n = b_n\Phi_n \tag{3.7.14}$$

showing that the Φ_n are simultaneously eigenfunctions of $\mathscr{A}$ and $\mathscr{B}$.

In the event that the eigenvalue a_n is d-fold degenerate, each $\mathscr{B}\Phi_{n\alpha}$ must evidently represent a linear combination of the degenerate eigenfunctions

$$\mathscr{B}\Phi_{n\alpha} = \sum_{\beta} c_\beta \Phi_{n\beta}. \tag{3.7.15}$$

However, by appropriate choice of the degenerate set, it is always possible to reduce each of (3.7.15) (for $\alpha = 1 \ldots d$) to diagonal form. Thereby, (3.7.14) will apply to every eigenfunction of $\mathscr{A}$, irrespective of degeneracy.

Two dynamical variables are *compatible* if their corresponding operators commute, in the sense that measurement of one will not perturb the observed values of the other. In terms of the generalized uncertainty relationship (3.6.21), commutativity of $\mathscr{A}$ and $\mathscr{B}$ implies that the uncertainty product $\Delta a \Delta b$ has zero as its lower limit. Zero is, in fact, attained whenever the system exists in one of the simultaneous eigenstates of $\mathscr{A}$ and $\mathscr{B}$.

C. *Dirac Notation*

In Dirac's formulation of quantum mechanics, the abstract vector associated with a wavefunction Ψ is denoted by the symbol $|\rangle$, called a *ket*. When Ψ is a member of some set of basis functions $\{\Phi_n\}$, most usually the eigenfunctions of some operator, the ket for Φ_n is written $|n\rangle$. The ket resulting from the action of a linear operator $\mathscr{A}$ on Φ_n is denoted $\mathscr{A}|n\rangle$. The eigenvalue equation (3.7.6) is accordingly written

$$\mathscr{A}|n\rangle = a_n|n\rangle, \tag{3.7.16}$$

or, for the continuum case,

$$\mathscr{A}|\nu\rangle = a(\nu)|\nu\rangle. \tag{3.7.17}$$

The result of multiplying Φ_n by the complex conjugate function Φ_m^* and integrating is formally equivalent to scalar premultiplication of $|n\rangle$ by the *bra*† vector $\langle m|$. This results in a further condensation of notation for integrals. viz.,

$$\langle m|n\rangle \equiv (\Phi_m, \Phi_n) = \int \Phi_m^* \Phi_n \, d\tau. \tag{3.7.18}$$

Analogously, one can represent integrals of the type (3.3.3) as *matrix elements* (cf. Section 3.11), viz.,

$$\langle m|\mathscr{A}|n\rangle \equiv (\Phi_m, \mathscr{A}\Phi_n) = \int \Phi_m^* \mathscr{A}\Phi_n \, d\tau. \tag{3.7.19}$$

The hermitian condition (3.3.5) thereby takes the form, in Dirac notation,

$$\langle m|\mathscr{A}|n\rangle = \langle n|\mathscr{A}|m\rangle^*. \tag{3.7.20}$$

3.8 Properties of Eigenstates

The hermiticity of a quantum-mechanical operator has important implications for its eigenstates. Consider first the discrete case, represented by the eigenvalue equation (3.7.16). Premultiplying by $\langle m|$, we obtain

$$\langle m|\mathscr{A}|n\rangle = a_n\langle m|n\rangle. \tag{3.8.1}$$

Interchanging m and n, then taking the complex conjugate, we find, in addition

$$\langle n|\mathscr{A}|m\rangle^* = a_m^*\langle n|m\rangle. \tag{3.8.2}$$

Now, by virtue of (3.7.20), the left-hand sides of (3.8.1) and (3.8.2) are equal. Therefore,

$$(a_n - a_m^*)\langle m|n\rangle = 0. \tag{3.8.3}$$

Consider first the case $m = n$. Since $\langle n|n\rangle \neq 0$ for a physically meaningful wavefunction (being, in fact, unity when Ψ_n is normalized), it is implied that

$$a_n^* = a_n. \tag{3.8.4}$$

The eigenvalues of a hermitian operator are thus real numbers. This is, in fact, quite necessary if, as presumed, the eigenvalues represent the possible results of measurement.

† Each bra $\langle n|$ represents the *dual vector* of the corresponding ket $|n\rangle$. A *bra* multiplied into a *ket* gives a *bracket* expression, enclosed by $\langle$ and $\rangle$.

When $m \neq n$ and $a_m \neq a_n$ (nondegenerate case), (3.8.3) implies that

$$\langle m|n\rangle = 0, \tag{3.8.5}$$

namely that nondegenerate eigenfunctions are orthogonal (cf. 3.2.10). Degenerate eigenfunctions need *not* be mutually orthogonal. The degenerate functions can always, however, be transformed by linear combination into a mutually orthogonal set—for example, by applying the Schmidt orthogonalization procedure.† Condition (3.8.5) will then apply to *all* pairs of eigenfunctions, irrespective of degeneracy. If, in addition, each eigenfunction is individually normalized, one can write

$$\langle m|n\rangle = \delta_{n,m} \tag{3.8.6}$$

in terms of the Kronecker delta (cf. A.6). A set of functions $\{\Phi_n\}$ obeying (3.8.6) is termed *orthonormal.*

From an alternative point of view, orthonormalization provides the most convenient solution to the equations

$$\sum_n a_n \langle m|n\rangle = a_m \sum_n \langle m|n\rangle \tag{3.8.7}$$

obtained by summing (3.8.3) (with $a_m^* = a_m$) over n. A *sufficient* (although not necessary) condition that (3.8.7) be fulfilled is that

$$\sum_n f_n \langle m|n\rangle = f_m \tag{3.8.8}$$

for arbitrary function f_n of the discrete variable n. Comparing (A.7), the obvious solution to (3.8.8) is the Kronecker orthonormalization condition (3.8.6).

For continuum eigenstates, it follows by analogous arguments that the eigenvalues are real and that nondegenerate eigenfunctions are orthogonal. Again the mutual orthogonality condition can be generalized for all distinct pairs of eigenfunctions, viz.,

$$\langle \mu|\nu\rangle = 0, \quad \text{if} \quad \mu \neq \nu. \tag{3.8.9}$$

† Suppose $\phi_1, \phi_2 \ldots \phi_d$, the original degenerate eigenfunctions, are not mutually orthogonal. Assume, however, that each has been normalized. An orthonormal set $\psi_1, \psi_2 \ldots \psi_d$ can be obtained by the following stepwise procedure. Let

$$\begin{aligned}
\psi_1 &= \phi_1 \\
\psi_2 &= \phi_2 - (\phi_2, \phi_1)\psi_1 \\
\psi_3 &= \phi_3 - (\phi_3, \phi_1)\psi_1 - (\phi_3, \phi_2)\psi_2 \\
&\ldots \\
\psi_i &= \phi_i - \sum_{j<i} (\phi_i, \phi_j)\psi_j, \qquad i = 1 \ldots d
\end{aligned}$$

It can be verified that the ith step results in a set of i mutually-orthogonal functions. As a final step, the new functions ψ_i are to be renormalized.

The notion of orthonormalization for continuum eigenfunctions requires further elaboration, however.

By analogous manipulation of the continuum eigenvalue equation (3.7.17), we arrive at

$$\int a(\nu)\langle\mu|\nu\rangle\, d\nu = a(\mu)\int \langle\mu|\nu\rangle\, d\nu \tag{3.8.10}$$

integration being over the eigenvalue continuum. This suggests, in analogy with (3.8.8), that the continuum overlap functions $\langle\mu|\nu\rangle$ satisfy

$$\int f(\nu)\langle\mu|\nu\rangle\, d\nu = f(\mu). \tag{3.8.11}$$

By virtue of (3.8.9), the integrand in (3.8.11) vanishes for every value of ν, *except* $\nu = \mu$. Since the integral does, nevertheless, have a finite value, the function $\langle\mu|\nu\rangle$ must evidently exhibit a strong singularity at the point $\nu = \mu$. Clearly, the requisite properties of $\langle\mu|\nu\rangle$ are possessed by the Dirac deltafunction $\delta(\nu - \mu)$ (cf. Appendix A). The general condition

$$\langle\mu|\nu\rangle = \delta(\nu - \mu) \tag{3.8.12}$$

is known as *deltafunction orthonormalization.* This is clearly the formal analogue of the Kronecker-delta orthonormalization condition (3.8.6). Free-particles eigenstates (cf. Chap. 5) provide an important illustration of deltafunction normalization. When ν, μ represent multiple quantum numbers, $\delta(\nu - \mu)$ is to be interpreted as a composite product of the type (A.36).

3.9 Eigenfunction Expansions

Quantum-mechanical eigenfunctions play an important formal role as basis functions in generalized Fourier expansions. The theory will first be developed for purely discrete basis sets, then generalized to include continuum functions.

Consider accordingly a set of basis functions $\{\phi_n(q)\}$ (eigenfunctions or otherwise) satisfying, for convenience, the Kronecker-delta orthonormality condition

$$(\phi_m, \phi_n) = \delta_{n,m}\,. \tag{3.9.1}$$

An arbitrary function $f(q)$ obeying the same analytic and boundary conditions as the basis functions can be initially approximated by a finite linear combination

$$f(q) \approx \sum_{n=1}^{N} c_n \phi_n(q), \tag{3.9.2}$$

the expansion coefficients c_n being, in general, complex numbers. It is convenient to determine these expansion coefficients such as to minimize the *error in the mean*:

$$\varepsilon_N \equiv \left\| f - \sum_{n=1}^{N} c_n \phi_n \right\|^2. \tag{3.9.3}$$

Expanding out the integral and using (3.9.1), we obtain

$$\varepsilon_N = (f, f) - \sum_n c_n (f, \phi_n) - \sum_n c_n^* (\phi_n, f) + \sum_n c_n^* c_n. \tag{3.9.4}$$

The error ε_N depends on 2N independent quantities: $c_1 \ldots c_N$ and $c_1^* \ldots c_N^*$.† For an absolute minimum in ε_N, it is necessary that

$$\frac{\partial \varepsilon_N}{\partial c_n} = 0, \quad \frac{\partial \varepsilon_N}{\partial c_n^*} = 0, \quad n = 1 \ldots N. \tag{3.9.5}$$

Applying (3.9.5), we obtain the conditions

$$\frac{\partial \varepsilon_N}{\partial c_n^*} = -(\phi_n, f) + c_n = 0, \tag{3.9.6}$$

along with the corresponding complex conjugates. The optimal values of the expansion coefficients are thereby given by

$$c_n = (\phi_n, f). \tag{3.9.7}$$

(This must certainly correspond to a *minimum* in ε_N since the possible error can be arbitrarily large.) Substitution into (3.9.4) gives the error in the mean for the N-term expansion;‡

$$\varepsilon_N = (f, f) - \sum_{n=1}^{N} |c_n|^2. \tag{3.9.8}$$

A crucial question concerns whether $\varepsilon_N \to 0$ as $N \to \infty$. If the expansion *does*, in fact, converge in the mean for arbitrary $f(q)$, the set of basis functions $\{\phi_n\}$ is said to be *complete*. §‖ We have, in that case,

$$(f, f) = \sum_{n=1}^{\infty} |c_n|^2, \tag{3.9.9}$$

† Alternatively, the real and imaginary parts of the c_n can be specified. The present choice leads more directly to the required result.

‡ An incidental result, since $\varepsilon_N \geqslant 0$, is that

$$(f, f) \geqslant \sum_{n=1}^{N} |c_n|^2,$$

known as Bessel's inequality.

§ A related concept is that of a *closed set* of functions. The set $\{\phi_n\}$ is closed if there exists *no* normalized function, within the given domain, which is orthogonal to *every* member of the set. Under very general conditions, closure implies completeness and vice versa.

‖ Problems of convergence do not, of course, arise if the basis set has a finite number of members or if $f(q)$ is exactly representable by a finite linear combination.

sometimes expressed

$$f(q) = \underset{N\to\infty}{\text{L.I.M.}} \sum_{n=1}^{N} c_n\phi_n(q). \tag{3.9.10}$$

If, in addition, the summation converges uniformly, the expansion actually represents the function, i.e.,†

$$f(q) = \sum_{n=1}^{\infty} c_n\Phi_n(q). \tag{3.9.11}$$

If the expansion theorem (3.9.11) is assumed at the outset, the values of the expansion coefficients, given by (3.9.7), follows by taking the scalar product of $f(q)$ with each $\phi_n(q)$ in turn and applying the orthonormality conditions (3.9.1).

For expansion of a function in terms of a complete set of *continuum* basis functions, the analog of (3.9.11) is clearly

$$f(q) = \int c(\nu)\phi_\nu(q)\,d\nu, \tag{3.9.12}$$

a generalization of the Fourier integral. Assuming deltafunction orthonormalization of the continuous basis, i.e.,

$$(\phi_\mu, \phi_\nu) = \delta(\nu - \mu) \tag{3.9.13}$$

it follows, analogous to (3.9.7), that

$$c(\nu) = (\phi_\nu, f). \tag{3.9.14}$$

For basis sets which contain *both* discrete and continuum functions, the expansion theorem takes the form

$$f(q) = \sum_n c_n\phi_n(q) + \int c(\nu)\phi_\nu(q)\,d\nu. \tag{3.9.15}$$

Discrete, continuous and composite basis sets have prototypes in the eigenfunctions of the harmonic oscillator, free particle and hydrogen atom, respectively.

The expansions (3.9.11), (3.9.12) and (3.9.15), appropriate to the respective spectral types can be represented by the following unified notation:

$$f(q) = \mathbf{S}_n\, c_n\phi_n(q). \tag{3.9.16}$$

The generalized summation symbol $\mathbf{S}_n$ calls for summation over discrete

† This result can alternatively be expressed in terms of the Fischer–Riesz theorem: Let $c_n \equiv (\phi_n, f)$, where $\{\phi_n\}$ is an orthonormal set and $f(q)$, an arbitrary function. Then, if the sum $\sum_{n=1}^{\infty} |c_n|^2$ converges, the expansion (3.9.11) is valid.

indices, integration over continuous spectral parameters, or the appropriate combination thereof.

Completeness of a basis set is compactly represented in Dirac notation as follows:

$$\mathop{\mathbf{S}}_{n} |n\rangle\langle n| = 1. \tag{3.9.17}$$

(This equation expresses the "resolution of the identity".) Postmultiplying (3.9.17) by an arbitrary ket $|\Phi\rangle$ and denoting

$$\langle n|\Psi\rangle = c_n, \tag{3.9.18}$$

we obtain

$$|\Psi\rangle = \mathop{\mathbf{S}}_{n} c_n |n\rangle, \tag{3.9.19}$$

which is the abstract vector equivalent of (3.9.16).

Proof of completeness must be carried out individually for every particular set of basis functions. For example, Fourier sine and cosine series can be shown to be complete with respect to expansion of simply-periodic piecewise-continuous functions. Although it has never been explicitly proven, it is generally assumed that every hermitian operator encountered in quantum mechanics possesses a complete set of eigenfunctions. This presumption, in fact, underlies much of the computational technique of quantum mechanics, including the variational principle and perturbation theory.

A. *Principle of Superposition*

The mathematical statement that the eigenfunctions of some quantum system comprise a complete set has as its physical counterpart the *principle of superposition*, whereby an arbitrary quantum state can be pictured as a superposition of eigenstates. In the notation of (3.9.16), this can be represented

$$\Psi(q, t) = \mathop{\mathbf{S}}_{n} c_n(t)\Phi_n(q, t). \tag{3.9.20}$$

A given wavefunction $\Psi(q, t)$ can, of course, be expanded in alternative ways, in terms of the eigenstates of different operators. Evaluating the norm of (3.9.20), it follows that

$$\mathop{\mathbf{S}}_{n} |c_n(t)|^2 = 1, \tag{3.9.21}$$

when Ψ is normalized and the Φ_n, appropriately orthonormalized. The last result can be interpreted in terms of composition of probability, wherein $|c_n|^2$ represents the weight or probability of eigenstate Φ_n in the wavefunction Ψ. In the continuum case, $|c(\nu)|^2$ correspondingly represents a probability density.

The expression for the expectation value $\langle \mathscr{A} \rangle$ (cf. 3.6.9), in which Ψ is expanded in the eigenfunctions of $\mathscr{A}$, transforms to

$$\langle \mathscr{A} \rangle = \mathop{\mathbf{S}}_{n} |c_n|^2 a_n. \tag{3.9.22}$$

This agrees with both (3.6.5) and (3.6.2), provided we identify $p_n = |c_n|^2$ in discrete ranges and $\rho(a) = |c(v)|^2$ in the continuum.

B. *Closure*

When the explicit forms of the coefficients, (3.9.7) and/or (3.9.14), are substituted, the expansion theorem (3.9.16) takes the form

$$f(q) = \mathop{\mathbf{S}}_{n} (\phi_n, f)\phi_n(q). \tag{3.9.23}$$

Now, writing out (ϕ_n, f) as an integral over primed dummy variables, and rearranging the integrand, we obtain

$$f(q) = \int f(q') \mathop{\mathbf{S}}_{n} \phi_n^*(q')\phi_n(q)\, \mathrm{d}\tau'. \tag{3.9.24}$$

Comparison with (A.1) shows that the generalized summation over n plays the role of deltafunction, viz.,

$$\mathop{\mathbf{S}}_{n} \phi_n^*(q')\phi_n(q) = \delta(q - q'). \tag{3.9.25}$$

This general property of eigenstates is known as *closure*. It is complementary to the orthonormalization property, (3.8.6) or (3.8.12), in the sense that the formal roles of the quantum numbers and configuration variables are interchanged.

By virtue of the closure property, integrals of the type $\langle m|\mathscr{A}|n\rangle$ (cf. Section 3.3) transform in the same way as matrix elements. To demonstrate this, consider the following calculation on the generalized summation $\mathbf{S}_k \langle m|\mathscr{A}|k\rangle\langle k|\mathscr{B}|n\rangle$. Writing out the integrals explicitly and making use of (3.9.25) and of integration over a delta-function, it follows that

$$\begin{aligned}
\mathop{\mathbf{S}}_{k} \langle m|\mathscr{A}|k\rangle\langle k|\mathscr{B}|n\rangle &= \iint \phi_m^*(q)\mathscr{A} \mathop{\mathbf{S}}_{k} \phi_k(q)\phi_k^*(q')\mathscr{B}'\phi_n(q')\, \mathrm{d}\tau\, \mathrm{d}\delta \\
&= \iint \phi_m^*(q)\mathscr{A}\delta(q - q')\mathscr{B}'\phi_n(q')\, \mathrm{d}\tau\, \mathrm{d}\tau' \\
&= \int \phi_m^*(q)\mathscr{A}\mathscr{B}\phi_n(q)\, \mathrm{d}\tau = \langle m|\mathscr{A}\mathscr{B}|n\rangle.
\end{aligned} \tag{3.9.26}$$

This result is isomorphic with the rule for multiplication of matrices, viz.,

$$\mathop{\mathbf{S}}_{k} A_{mk}B_{kn} = (AB)_{mn}. \tag{3.9.27}$$

It is accordingly convenient to utilize the notation

$$A_{mn} \equiv \langle m|\mathscr{A}|n\rangle, \tag{3.9.28}$$

identifying such operator integrals as *matrix elements*. In Section 3.11, we shall develop more systematically the connection between operators and matrices.

3.10 Hilbert Space

Certain formal correspondences between the theories of vector spaces and function spaces enable the formalism of quantum mechanics to be given an abstract geometrical interpretation. To develop this representation, let us note some obvious analogies between formulas of vector analysis and those of function theory—with particular reference to the generalized Fourier expansions of functions discussed in the preceding section.

Define an operation of *inner multiplication* as follows. For two vectors the inner product stands for the scalar product, i.e.,

$$(a, b) \equiv \mathbf{a} \cdot \mathbf{b}. \tag{3.10.1$'$}$$

For two functions, it stands for the complex overlap integral, i.e.,

$$(f, g) \equiv \int f^*(q) g(q)\, d\tau. \tag{3.10.1}$$

Three-dimensional vector space is spanned by a set of three orthonormal base (or unit) vectors $\mathbf{e}_1$, $\mathbf{e}_2$, $\mathbf{e}_3$ (more commonly denoted **i**, **j**, **k**) whose scalar products satisfy conditions

$$(e_i, e_j) = \delta_{ij}. \tag{3.10.2$'$}$$

A given function domain can likewise be spanned by an orthonormal set of basis functions $\{\phi_n(q)\}$, wherein

$$(\phi_m, \phi_n) = \delta_{mn} \tag{3.10.2}$$

(or the corresponding deltafunction). When the number of basis functions is infinite, the associated *function space* will likewise be infinite-dimensional.

An arbitrary vector **a** can be represented

$$\mathbf{a} = \sum_{i=1}^{3} a_i \mathbf{e}_i \tag{3.10.3$'$}$$

in terms of its cartesian components a_1, a_2, a_3 along the three basic axes. These components can be formally represented by

$$a_i = (e_i, a), \qquad i = 1, 2, 3. \tag{3.10.4$'$}$$

An arbitrary function $f(q)$ can be represented by an expansion (cf. 3.9.16)

$$f(q) = \mathbf{S}_n c_n \phi_n(q) \tag{3.10.3}$$

in which the expansion coefficients are given by

$$c_n = (\phi_n, f). \tag{3.10.4}$$

The scalar product of two vectors is given, in terms of their components, by

$$(a, b) = \mathbf{a} \cdot \mathbf{b} = \sum_{i=1}^{3} a_i b_i. \tag{3.10.5'}$$

The norm or length of a vector is accordingly

$$\|a\| = (a, a)^{\frac{1}{2}} = \left[\sum_{i=1}^{3} a_i^2\right]^{\frac{1}{2}}. \tag{3.10.6'}$$

If two functions $f(q)$ and $g(q)$ are expanded in terms of the same orthonormal basis set, their inner product is given by

$$(f, g) = \mathbf{S}_n c_n^* c_n' \tag{3.10.5}$$

where c_n and c_n' are the expansion coefficients for $f(q)$ and $g(q)$, respectively. The norm of a function (cf. 3.2.9) is given, in terms of its expansion coefficients, by

$$\|f\| = (f, f)^{\frac{1}{2}} = [\mathbf{S}_n |c_n|^2]^{\frac{1}{2}}. \tag{3.10.6}$$

It should now be clear, in view of the analogies between the preceding primed and unprimed equations, that functions can be conceptualized as vectors in an infinite-dimensional space. The expansion coefficients c_n with respect to a basis $\{\phi_n(q)\}$ play the role of cartesian coordinates for these abstract vectors. The particular function space which admits of complex normalizable functions is known as *Hilbert space*.† Complex values for the expansion coefficients are admissible, which accounts for the generalization of the inner product in conformity with (3.10.1).

The state of a quantum system is represented by a vector (more generally, a direction) in Hilbert space. Dirac's bra and ket notation (cf. Section 3.7) pertains, in fact, to these state vectors. Correspondingly, every operator is represented by a linear transformation among Hilbert-space vectors.

† Strictly speaking, Hilbert space is limited to normalizable functions, for which $\sum_{n=1}^{\infty} |c_n|^2$ converges. The theory would accordingly be limited to eigenfunctions belonging to discrete spectra, (at most) denumerably infinite in number. For quantum-theoretical purposes, one can, however, conceptualize a generalization of Hilbert space, amenable, as well, to deltafunction-normalizable functions and to non-denumerable basis sets.

3.11 Matrices in Quantum Mechanics

Relationships among three-dimensional vectors and linear transformations (rotations, reflections, etc.) can be usefully represented in terms of matrices.† Matrix notation for vectors and operators in Hilbert space is likewise an important adjunct to quantum-mechanical computational technique.

Consider accordingly an operator equation such as

$$g = \mathscr{A}f. \tag{3.11.1}$$

Reexpressed in terms of the expansions $f(q) = \mathbf{S}_n c_n\phi_n(q)$ and $g(q) = \mathbf{S}_n c'_n\phi_n(q)$ this becomes

$$\mathbf{S}_n c'_n\phi_n(q) = \mathbf{S}_n c_n\mathscr{A}\phi_n(q), \tag{3.11.2}$$

where use is made of the presumed linearity of $\mathscr{A}$ (cf. 3.3.2). Taking successively the inner product with each $\phi_k(q)$, we obtain a set of equations for the vector components, viz.,

$$c'_k = \mathbf{S}_n A_{kn}c_n, \tag{3.11.3}$$

where matrix notation (cf. 3.9.28) has been introduced for the transformation coefficients.

If, by a subsequent linear transformation, we obtain

$$h = \mathscr{B}g, \tag{3.11.4}$$

where $h(q) = \mathbf{S}_n c''_n\phi_n(q)$, it must also be possible to effect a direct transformation of $f(q)$ into $h(q)$, viz.,

$$h = \mathscr{C}f. \tag{3.11.5}$$

Writing down the appropriate transformation equations, we have

$$c''_m = \mathbf{S}_k B_{mk}c'_k = \mathbf{S}_n\mathbf{S}_k B_{mk}A_{kn}c_n, \tag{3.11.6}$$

while from (3.11.5)

$$c''_m = \mathbf{S}_n C_{mn}c_n. \tag{3.11.7}$$

Comparing the last two equations, it is evident that an operator product

$$\mathscr{C} = \mathscr{B}\mathscr{A} \tag{3.11.8}$$

has associated transformation coefficients related by

$$C_{mn} = \mathbf{S}_k B_{mk}A_{kn}\,. \tag{3.11.9}$$

† For a review of matrix algebra, see, for example, Margenau and Murphy, "Mathematics of Physics and Chemistry", Van Nostrand, New York, 1956, Chapter 10.

We arrive thus at the fundamental rule for matrix multiplication, already demonstrated by eqn (3.9.27).

An alternative representation of a linear operator $\mathscr{A}$ can be based on the characterization of its action on any member of a given basis set $\{\phi_n\}$ according to

$$\mathscr{A}\phi_n = \mathbf{S}_m \phi_m A_{mn}. \tag{3.11.10}$$

Every basis function is thereby uniquely transformed by action of an operator into a linear combination whose coefficients are the elements of the matrix A_{mn}. The integral representation of an operator (3.3.7) is actually a special case of (3.11.10), the kernel $A(q, q')$ constituting, in essence, a continuous matrix.

It is convenient that this point to introduce matrix notation explicitly. Every vector can be symbolized by a one-column array of its coefficients, thus:

$$\mathbb{C} = \begin{bmatrix} c_1 \\ c_2 \\ c_3 \end{bmatrix} \tag{3.11.11}$$

Every linear operator can correspondingly be represented by a square array of its transformation coefficients, thus:

$$\mathbb{A} = \begin{bmatrix} A_{11} & A_{12} & A_{13} & \cdots \\ A_{21} & A_{22} & A_{23} & \cdots \\ A_{31} & A_{32} & A_{33} & \cdots \\ \cdots & \cdots & \cdots & \cdots \end{bmatrix} \tag{3.11.12}$$

These arrays, or *matrices*, obey the same algebraic relations as the corresponding vectors and operators. For example,

$$\mathbb{c}' = \mathbb{A}\mathbb{c} \tag{3.11.13}$$

corresponds to (3.11.1) and

$$\mathbb{C} = \mathbb{B}\mathbb{A} \tag{3.11.14}$$

to (3.11.8). Matrix multiplication symbolizes a prescribed set of combinatorial relations among matrix elements. Accordingly (3.11.13) and (3.11.14) represent the sets of transformation equations (3.11.3) and (3.11.9), respectively.

When the basis set $\{\phi_n\}$ is denumerably or nondenumerably infinite, the matrices representing quantum-mechanical states and observables are correspondingly of infinite rank. Most of the results pertaining to finite matrices can nevertheless be taken over, one stipulation being that summations for matrix products converge appropriately.†

A hermitian operator—hence every quantum-mechanical observable—can be represented by a hermitian matrix, whose elements, given by (3.9.28), satisfy the conditions

$$A_{mn} = A_{nm}^*. \tag{3.11.15}$$

There exist, however, arbitrarily many matrix representations for every operator, depending upon the particular choice of basis functions $\{\phi_n\}$. These alternative representations are related to one another by unitary transformations, whereby

$$\mathbb{A}' = \mathbb{U}^\dagger \mathbb{A} \mathbb{U} \tag{3.11.16}$$

Every hermitian matrix may, in one such representation, be reduced to diagonal form,‡ viz.,

$$\mathbb{A}' = \begin{bmatrix} a_1 & 0 & 0 & \cdots \\ 0 & a_2 & 0 & \cdots \\ 0 & 0 & a_3 & \cdots \\ \cdots & \cdots & \cdots & \cdots \end{bmatrix}, \tag{3.11.17}$$

thereby exhibiting its eigenvalues. For a hermitian matrix, these are real numbers. Diagonalization of a matrix is mathematically equivalent to solving the eigenvalue equation

$$\mathbb{A}\mathbb{c}_n = a_n \mathbb{c}_n, \tag{3.11.18}$$

the matrix analog of (3.7.6).

Given a matrix $\mathbb{A}$ in a nondiagonal representation, the eigenvalues can be found by solving the secular equation

$$\begin{vmatrix} A_{11} - a & A_{12} & A_{13} & \cdots \\ A_{21} & A_{22} - a & A_{23} & \cdots \\ A_{31} & A_{32} & A_{33} - a & \cdots \\ \cdots & \cdots & \cdots & \cdots \end{vmatrix} = 0 \tag{3.11.19}$$

The same equation can be utilized in linear variational calculations, in which the matrix elements are evaluated w.r.t. some *incomplete* basis set. The roots of the secular determinant accordingly represent approximations to the true eigenvalues.§

† For more detailed discussions of infinite matrices, see J. von Neumann, *Mathematical Foundations of Quantum Mechanics* (Princeton Univ. Press, 1955).

‡ There are subtle points involved as to whether every infinite hermitian matrix can be diagonalized. See J. von Neumann, *op cit.*

§ Various generalizations on the interleaving of approximate and exact eigenvalues have been derived by J. K. L. MacDonald, *Phys. Rev.* **43**, 830 (1933).

4

Principles of Quantum Dynamics

4.1 Time-dependent Schrödinger Equation

Postulate 5. The time development of a quantum-mechanical system is governed by the time-dependent Schrödinger equation

$$i\hbar \frac{\partial \Psi}{\partial t} = \mathscr{H}\Psi. \tag{4.1.1}$$

Formally this equation derives from quantization of the fourth component of the momentum 4-vector, i.e.,

$$p_4 \to -i\hbar\,\partial/\partial x_4\,. \tag{4.1.2}$$

Making use of (2.4.2) and (2.5.28), this is equivalent to

$$\frac{iE}{c} \to -\frac{i\hbar}{ic}\frac{\partial}{\partial t} \tag{4.1.3}$$

hence to

$$E \to i\hbar\,\partial/\partial t. \tag{4.1.4}$$

The time-dependent Schrödinger equation is then arrived at by 4-dimensional quantization of the classical relation

$$H(q, p, t) = E. \tag{4.1.5}$$

For a particle in a conservative field, the time-dependent Schrödinger equation takes the explicit form

$$i\hbar \frac{\partial}{\partial t}\Psi(\mathbf{r}, t) = \left\{-\frac{\hbar^2}{2m}\nabla^2 + V(\mathbf{r})\right\}\Psi(\mathbf{r}, t). \tag{4.1.6}$$

This is of course based on quantization of the *nonrelativistic* energy–momentum relation

$$E = p^2/2m + V(\mathbf{r}). \tag{4.1.7}$$

The nonrelativistic character of (4.1.6) is most obvious from the lack of symmetry between space and time derivatives. A relativistic wave equation can be arrived at by quantization of the relativistic energy-momentum relation (2.5.30):

$$E^2 = p^2c^2 + m^2c^4. \tag{4.1.8}$$

For the free particle this gives

$$\frac{1}{c^2}\frac{\partial^2\Psi}{\partial t^2} = \nabla^2\Psi - \left(\frac{mc}{\hbar}\right)^2 \Psi, \tag{4.1.9}$$

which is known as the Klein–Gordon equation. It is valid, however, for zero-spin particles such as pions. The relativistic quantum theory of the electron is governed by the Dirac equation

$$i\hbar\, \partial\Psi/\partial t = (-i\hbar\alpha \cdot \nabla + \beta mc^2)\Psi, \tag{4.1.10}$$

which is of first order in both space and time derivatives.

4.2 Connection with Hamilton–Jacobi Theory

The time-dependent Schrödinger equation (4.1.1) and the Hamilton–Jacobi equation (1.4.16):

$$\frac{\partial S}{\partial t} + H\left(q, \frac{\partial S}{\partial q}, t\right) = 0 \tag{4.2.1}$$

are closely analogous in structure: each relates a Hamiltonian to a time derivative. Schrödinger's original postulation was in fact, based on this analogy.

For concreteness, we shall consider the case of a particle in a conservative field, but our conclusions will apply more generally. Corresponding to the Schrödinger equation (4.1.6) is the Hamilton–Jacobi equation

$$\frac{\partial S}{\partial t} + \frac{1}{2m}(\nabla S)^2 + V(\mathbf{r}) = 0. \tag{4.2.2}$$

To develop the connection, let the solution to (4.1.6) be represented in the form

$$\Psi(\mathbf{r}, t) = \text{const} \exp\left[(i/\hbar)\,\Phi(\mathbf{r}, t)\right]. \tag{4.2.3}$$

The Schrödinger equation thereby transforms to

$$\left[\frac{\partial \Phi}{\partial t} + \frac{1}{2m}(\nabla\Phi)^2 - \frac{i\hbar}{2m}\nabla^2\Phi + V(\mathbf{r})\right]\Psi = 0. \tag{4.2.4}$$

Except for the third term, we should have obtained (4.2.2). This suggests that the phase function $\Phi(\mathbf{r}, t)$ be represented as a power series in $\hbar$:

$$\Phi = S + \hbar S_1 + \hbar^2 S_2 + \dots \tag{4.2.5}$$

in which S is Hamilton's principal function (the solution to 4.2.2). We obtain thereby

$$\left[\frac{\partial S}{\partial t} + \frac{1}{2m}(\nabla S)^2 + V(\mathbf{r})\right]\Psi + O(\hbar) + \dots = 0. \tag{4.2.6}$$

Evidently, in the classical limit $\hbar \to 0$, the time-dependent Schrödinger equation becomes equivalent to the Hamilton–Jacobi equation, their respective solutions being related by:†

$$\Psi \to \text{const} \exp\left[(i/\hbar)\, S\right] \qquad \text{as } \hbar \to 0. \tag{4.2.7}$$

This limiting relationship suggests that the actual wavefunction for a quantum system can be represented in the form

$$\Psi(q, t) = F(q, t) \exp\left[(i/\hbar)\, S(q, t)\right] \tag{4.2.8}$$

in which the amplitude function $F(q, t)$ takes account of the residual terms in (4.2.5). Different quantum states of a given system will be characterized by different amplitude functions but they will have in common the exponential dependence on Hamilton's principal function. This is a useful adjunct to the construction of Green's functions (cf. Section 6.5).

A correspondence exists as well between the time-independent Schrödinger and Hamilton–Jacobi equations. In the limit $\hbar \to 0$,

$$\mathscr{H}\psi(q) = E\psi(q) \tag{4.2.9}$$

reduces to

$$H\left(q, \frac{\partial W}{\partial q}\right) = E \tag{4.2.10}$$

in such way that

$$\psi(q) \to \text{const} \exp\left[(i/\hbar)\, W(q)\right] \text{as } \hbar \to 0 \tag{4.2.11}$$

A more accurate approximation to the wavefunction is provided by the WKB method, in which terms up to first order in $\hbar$ are accounted for.

4.3 Permanence of Normalization

Every solution of the time-dependent Schrödinger equation, once normalized, maintains its normalization in time. This is shown by the following

† For the free particle, $\Psi(\mathbf{r}, t) = \exp\left[(i/\hbar)(\mathbf{p}\cdot\mathbf{r} - Et)\right]$ while $S(\mathbf{r}, t) = \mathbf{p}\cdot\mathbf{r} - Et$, so that (4.2.7) is the exact solution.

calculation. First,†

$$\frac{\mathrm{d}}{\mathrm{d}t}(\Psi, \Psi) = (\partial\Psi/\partial t, \Psi) + (\Psi, \partial\Psi/\partial t). \tag{4.3.1}$$

Now make use, in the two integrals, of (4.1.1) and its complex conjugate equation, viz.,

$$-i\hbar\, \partial\Psi^*/\partial t = (\mathscr{H}\Psi)^*. \tag{4.3.2}$$

The right-hand side becomes

$$(i\hbar)^{-1}[-(\mathscr{H}\Psi, \Psi) + (\Psi, \mathscr{H}\Psi)], \tag{4.3.3}$$

which vanishes by virtue of the hermitian property of the Hamiltonian. Therefore

$$\frac{\mathrm{d}}{\mathrm{d}t}(\Psi, \Psi) = 0. \tag{4.3.4}$$

4.4 Continuity Equation

Premultiplying (4.1.6), the time-dependent Schrödinger equation for a particle, by Ψ^* and subtracting from this the corresponding complex conjugate equation, there results

$$i\hbar\left(\Psi^*\frac{\partial\Psi}{\partial t} + \frac{\partial\Psi^*}{\partial t}\Psi\right) = -\frac{\hbar^2}{2m}(\Psi^*\nabla^2\Psi - \Psi\nabla^2\Psi^*). \tag{4.4.1}$$

This can be rearranged to

$$\frac{\partial}{\partial t}(\Psi^*\Psi) - \frac{i\hbar}{2m}\nabla\cdot(\Psi^*\nabla\Psi - \Psi\nabla\Psi^*) = 0, \tag{4.4.2}$$

which has the form of the equation of continuity (cf. 2.1.8)

$$\frac{\partial\rho}{\partial t} + \nabla\cdot\mathbf{j} = 0. \tag{4.4.3}$$

In its quantum-mechanical application the continuity equation (4.4.2), evidently pertains to a hypothetical probability fluid. The density function is identified as

$$\rho(\mathbf{r}, t) = \Psi^*(\mathbf{r}, t)\,\Psi(\mathbf{r}, t) \tag{4.4.4}$$

† As to use of the partial and total derivative w.r.t. t, note that

$$\frac{\mathrm{d}}{\mathrm{d}t}\int f(q, t)\,\mathrm{d}\tau = \int \frac{\partial f}{\partial t}\,\mathrm{d}\tau\,.$$

in agreement with Born's and Schrödinger's interpretation of the wavefunction in terms of particle probability density (cf. Section 3.2 and eqn 3.2.1). The probability current density—flux of probability across unit area per unit time—is correspondingly represented by†

$$\mathbf{j}(\mathbf{r}, t) = -\frac{i\hbar}{2m}(\Psi^*\nabla\Psi - \Psi\nabla\Psi^*). \tag{4.4.5}$$

The densities (4.4.4) and (4.4.5), when multiplied by e, can be given somewhat more physical characterizations as electrical charge and current densities, respectively. Similarly when multiplied by m, they represent, respectively, mass density and mass flux density—the latter being equivalent to momentum density.

Introducing into (4.4.5) the "velocity operator", $\mathbf{v} = \mathbf{p}/m = -(i\hbar/m)\nabla$, the probability current density can be expressed

$$\mathbf{j} = \mathrm{Re}\,\Psi^*\mathbf{v}\Psi. \tag{4.4.6}$$

This is in essential agreement with the classical formula pertaining to convective flow (cf. 2.1.1):

$$\mathbf{j} = \rho\mathbf{v}. \tag{4.4.7}$$

4.5 Heisenberg's Equation of Motion

Consider a matrix element

$$A_{mn} = (\Psi_m, \mathscr{A}\Psi_n) \tag{4.5.1}$$

in which Ψ_m and Ψ_n are two wavefunctions obeying the time-dependent Schrödinger equation. The time dependence of A_{mn} is deduced in the following calculation. Differentiating wrt t,

$$\frac{\mathrm{d}A_{mn}}{\mathrm{d}t} = \left(\frac{\partial\Psi_m}{\partial t}, \mathscr{A}\Psi_n\right) + \left(\Psi_m, \frac{\partial\mathscr{A}}{\partial t}\Psi_n\right) + \left(\Psi_m, \mathscr{A}\frac{\partial\Psi_n}{\partial t}\right). \tag{4.5.2}$$

The second integral reflects any explicit time dependence of the operator $\mathscr{A}$. By virtue of (4.1.1) and (4.3.4), the third integral equals $(i\hbar)^{-1}(\Psi_m, \mathscr{A}\mathscr{H}\Psi_n)$ while the first equals $-(i\hbar)^{-1}(\mathscr{H}\Psi_m, \mathscr{A}\Psi_n)$. But by the hermitian property of $\mathscr{H}$,

$$(\mathscr{H}\Psi_m, \mathscr{A}\Psi_n) = (\Psi_m, \mathscr{H}\mathscr{A}\Psi_n). \tag{4.5.3}$$

In terms of the commutator of $\mathscr{A}$ and $\mathscr{H}$ (cf. 3.3.13), we obtain therefore

$$\frac{\mathrm{d}A_{mn}}{\mathrm{d}t} = \left(\Psi_m, \frac{\partial\mathscr{A}}{\partial t}\Psi_n\right) + (i\hbar)^{-1}(\Psi_m, [\mathscr{A}, \mathscr{H}]\Psi_n). \tag{4.5.4}$$

† There is an additional term in the presence of a vector potential (cf. Section 8.3).

Introducing matrix notation for the integrals, we obtain *Heisenberg's equation of motion*:

$$\frac{\mathrm{d}A_{mn}}{\mathrm{d}t} = \left(\frac{\partial A}{\partial t}\right)_{mn} + (i\hbar)^{-1}[A, H]_{mn}. \tag{4.5.5}$$

Expressed as an abstract matrix equation:

$$\frac{\mathrm{d}\mathbb{A}}{\mathrm{d}t} = \frac{\partial \mathbb{A}}{\partial t} + (i\hbar)^{-1}(\mathbb{A}\mathbb{H} - \mathbb{H}\mathbb{A}), \tag{4.5.6}$$

this represents the fundamental dynamical principle of matrix mechanics (cf. Section 4.14).

The time-dependence of quantum-mechanical expectation values is found by specialization of the Heisenberg equation of motion—e.g., (4.5.4)—to the case $\Psi_m = \Psi_n = \Psi$. Dividing through by (Ψ, Ψ)—which, according to (4.3.4), is time-independent—and using the notation of eqn (3.6.9),

$$\frac{\mathrm{d}}{\mathrm{d}t}\langle \mathscr{A} \rangle = \left\langle \frac{\partial \mathscr{A}}{\partial t} \right\rangle + (i\hbar)^{-1}\langle [\mathscr{A}, \mathscr{H}] \rangle. \tag{4.5.7}$$

The expectation value of dynamical variable is independent of time if the corresponding operator (i) contains no explicit dependence on time, i.e.,

$$\partial \mathscr{A}/\partial t = 0 \tag{4.5.8}$$

and (ii) commutes with the Hamiltonian,† i.e.

$$[\mathscr{A}, \mathscr{H}] = 0. \tag{4.5.9}$$

The latter condition is usually possible only if $\mathscr{H}$ is also time-independent. The expectation value $\langle \mathscr{A} \rangle$ of an operator fulfilling (4.5.8) and (4.5.9) is known as a *constant of the motion*, the terminology being borrowed from classical mechanics. A constant of the motion is, most often, a generalized coordinate or momentum variable whose canonical conjugate does not appear in the Hamiltonian. Two important examples are linear momentum, which is conserved when $\mathscr{H}$ is translationally invariant, and angular momentum, which is conserved when $\mathscr{H}$ has spherical symmetry. On a more fundamental level, every symmetry in nature gives rise to a conservation law.

When applied to the Hamiltonian itself, the Heisenberg equation of motion (4.5.7) reads

$$\frac{\mathrm{d}}{\mathrm{d}t}\langle \mathscr{H} \rangle = \left\langle \frac{\partial \mathscr{H}}{\partial t} \right\rangle, \tag{4.5.10}$$

† Recall that this is also the condition for existence of simultaneous eigenfunctions of $\mathscr{A}$ and $\mathscr{H}$ (cf. Section 3.7B).

showing that time-dependence in $\mathscr{H}$ can arise only from explicit dependence of the Hamiltonian on t. For conservative dynamical systems, the Hamiltonian is time-independent:

$$\partial\mathscr{H}/\partial t = 0. \tag{4.5.11}$$

Its expectation value becomes accordingly a constant of the motion,

$$\langle\mathscr{H}\rangle = E \tag{4.5.12}$$

namely, the energy of the system. Stated another way, time-displacement invariance in a system implies energy conservation.

4.6 Ehrenfest's Theorem

Expectation values of generalized coordinates and momenta obey the equations of motion

$$\frac{d\langle q_i\rangle}{dt} = (i\hbar)^{-1}\langle[q_i, \mathscr{H}]\rangle \tag{4.6.1}$$

and

$$\frac{d\langle p_i\rangle}{dt} = (i\hbar)^{-1}\langle[p_i, \mathscr{H}]\rangle. \tag{4.6.2}$$

A system of particles in a conservative potential field can be represented by the Hamiltonian

$$\mathscr{H} = \sum_{k=1}^{N}\frac{p_k^2}{2m_k} + V(\mathbf{r}_1 \ldots \mathbf{r}_N), \tag{4.6.3}$$

where (cf. 3.4.11) $\mathbf{p}_k = -i\hbar\nabla_k$. The operator

$$\mathbf{F}_k \equiv -\nabla_k V(\mathbf{r}_1 \ldots \mathbf{r}_N) \tag{4.6.4}$$

corresponds to the classical force on particle k. With the aid of (3.3.17), (3.4.4) and (3.4.5), we deduce the commutation relations

$$[\mathbf{r}_k, \mathscr{H}] = \frac{i\hbar}{m}\mathbf{p}_k \tag{4.6.5}$$

and

$$[\mathbf{p}_k, \mathscr{H}] = i\hbar\mathbf{F}_k. \tag{4.6.6}$$

Substituting these into (4.6.1) and (4.6.2), respectively, we obtain

$$\frac{d}{dt}\langle\mathbf{r}_k\rangle = \frac{\langle\mathbf{p}_k\rangle}{m} \tag{4.6.7}$$

and

$$\frac{\mathrm{d}}{\mathrm{d}t}\langle \mathbf{p}_k \rangle = \langle \mathbf{F}_k \rangle \tag{4.6.8}$$

which are called *Ehrenfest's relations.* These are closely analogous to the corresponding classical equations of motion. By differentiating (4.6.7) wrt t and using (4.6.8) for $d\langle \mathbf{p}_k \rangle/\mathrm{d}t$, we obtain

$$\langle \mathbf{F}_k \rangle = m\frac{\mathrm{d}^2}{\mathrm{d}t^2}\langle \mathbf{r}_k \rangle, \tag{4.6.9}$$

which is clearly the quantum-mechanical analog of Newton's second law. The physical content of this relation is known as *Ehrenfest's theorem,*† namely that expectation values of position and momentum coordinates conform to the classical laws of motion. Applied to a single particle, this means specifically that the centroid of the associated wavepacket follows the classical trajectory.

The limiting validity of classical mechanics in a quantum universe rests, in fact, on Ehrenfest's theorem. Classical mechanics provides an accurate—or, at least, useful—description of the motions of particles whenever the spreads of their wavepackets are small compared to their parameters of motion.

4.7 Time Inversion Symmetry

The hypothetical behaviour of a quantum system under the transformation

$$t \to -t, \tag{4.7.1}$$

known as *time reversal* or *time inversion*, is of fundamental significance in elementary particle theory.§ Applying (4.7.1) to (4.1.1) we obtain

$$-i\hbar\frac{\partial}{\partial t}\Psi(q, -t) = \mathscr{H}(-t)\Psi(q, -t). \tag{4.7.2}$$

Comparing this with the complex conjugate equation (4.3.2), it is seen that if the Hamiltonian fulfils the condition

$$\mathscr{H}(-t) = \mathscr{H}^*(t), \tag{4.7.3}$$

the wavefunction transforms according to

$$\Psi(q, -t) = \Psi^*(q, t). \tag{4.7.4}$$

† P. Ehrenfest, *Z. Phys.* **45**, 455 (1927).

§ Time inversion belongs to the trio of fundamental symmetry operations designated "CPT". The other two are charged conjugation ($e \to -e$) and parity ($\mathbf{r} \to -\mathbf{r}$).

The last relation would imply that the physical behaviour of the system is invariant under time reversal since

$$|\Psi(q, -t)|^2 = |\Psi(q, t)|^2, \tag{4.7.5}$$

contingent of course, on the validity of (4.7.3). For time-independent Hamiltonian, time-reversal invariance evidently requires that $\mathscr{H}$ be *real* as well as hermitian ($\mathscr{H}^* = \mathscr{H}$).

Time-inversion symmetry also occurs in classical dynamics. For conservative systems, Newton's second law is invariant under (4.7.1) since it involves the second time derivative. More generally under the condition that

$$L(q, -\dot{q}, -t) = L(q, \dot{q}, t) \tag{4.7.6}$$

Lagrange's equations (1.1.8) remain valid if when each $q_i(-t) = q_i(t)$ and each $\dot{q}_i(-t) = -\dot{q}_i(t)$. The last relation also implies that $p_i(-t) = -p_i(t)$. In the Hamiltonian formulation, the corresponding condition for time-inversion symmetry is that

$$H(q, -p, -t) = H(q, p, t). \tag{4.7.7}$$

Since complex conjugation changes the sign of quantum-mechanical momentum operators $-i\hbar\, \partial/\partial q_i$, (4.7.3) in equivalent to (4.7.7).

Angular momentum operators, including spins, also change sign under time inversion. An interesting consequence of this is the *Kramers degeneracy*, wherein every quantum system having an odd number of spins $\frac{1}{2}$ has at least a twofold degeneracy. This is implied by the fact that $\Psi(q, t)$ and $\Psi^*(q, t)$ will represent *distinct* solutions of the Schrödinger equation.

Time-inversion symmetry always applies for conservative systems when the Hamiltonian is a homogeneous quadratic function of the momentum variables. This is preserved in a static electric field but destroyed by a magnetic field, which introduces coupling terms linear in momentum (cf. Section 8.1). The Kramers degeneracy is thus resolved in the latter case.

The primary physical laws governing individual particles are characterized by time invariance, in both classical and quantum mechanics. This is the *principle of microscopic reversibility*. However, in secondary principles involving the statistical behaviour of large assemblies of particles—such as the second law of thermodynamics—probabilistic considerations destroy such invariance, thus creating, in Eddington's words, the "arrow of time". Still, for systems at equilibrium, the rate of every molecular process is exactly equal to the rate of the inverse process (*principle of detailed balancing*).

4.8 Formal Solution of the Time-dependent Schrödinger Equation

Since the time-dependent Schrödinger equation (4.1.1) is a first-order differential equation in time, an initial condition $\Psi(q, t_0)$ should suffice, in principle, to determine the wavefunction for all subsequent (and past) times. To develop this general solution, integrate (4.1.1) between t_0 and t. This results in an integral equation

$$\Psi(q, t) = \Psi(q, t_0) + (i\hbar)^{-1} \int_{t_0}^{t} \mathscr{H}(t')\,\Psi(q, t')\,\mathrm{d}t'. \tag{4.8.1}$$

Here t' is simply the dummy variable of integration, with $t_0 \leqslant t' \leqslant t$. Changing variables in (4.8.1), whereby $t' \to t''$, $t \to t'$, we obtain

$$\Psi(q, t') = \Psi(q, t_0) + (i\hbar)^{-1} \int_{t_0}^{t'} \mathscr{H}(t'')\,\Psi(q, t'')\,\mathrm{d}t'', \tag{4.8.2}$$

consistent with $t_0 \leqslant t'' \leqslant t'$. Substituting (4.8.2) into (4.8.1),

$$\Psi(q, t) = \Psi(q, t_0) + (i\hbar)^{-1} \int_{t_0}^{t} \mathscr{H}(t')\,\Psi(q, t_0)\,\mathrm{d}t' + (i\hbar)^{-2} \int_{t_0}^{t} \int_{t_0}^{t'} \mathscr{H}(t')\,\mathscr{H}(t'')\,\Psi(q, t'')\,\mathrm{d}t'\,\mathrm{d}t''. \tag{4.8.3}$$

Continuing this iterative procedure, we arrive finally at

$$\Psi(q, t) = \mathscr{U}(t, t_0)\,\Psi(q, t_0), \tag{4.8.4}$$

in terms of the *evolution operator* (or *time-development operator*)

$$\mathscr{U}(t, t_0) \equiv 1 + (i\hbar)^{-1} \int_{t_0}^{t} \mathscr{H}(t')\,\mathrm{d}t' + (i\hbar)^{-2} \int_{t_0}^{t} \int_{t_0}^{t'} \mathscr{H}(t')\,\mathscr{H}(t'')\,\mathrm{d}t'\,\mathrm{d}t'' + \ldots \tag{4.8.5}$$

This time-integral expansion can be written

$$\mathscr{U}(t, t_0) = 1 + \sum_{n=1}^{\infty} (i\hbar)^{-n} \int_{t_0}^{t} \int_{t_0}^{t'} \cdots \int_{t_0}^{t^{(n-1)}} \times \mathscr{H}(t')\mathscr{H}(t'')\ldots\mathscr{H}(t^{(n)})\,\mathrm{d}t'\,\mathrm{d}t''\ldots\mathrm{d}t^{(n)}. \tag{4.8.6}$$

The limits of integration imply the time sequence

$$t \geqslant t' \geqslant t'' \geqslant t''' \geqslant \ldots \geqslant t_0\,. \tag{4.8.7}$$

The operator product $\mathscr{H}(t')\,\mathscr{H}(t'')\ldots\mathscr{H}(t^{(n)})$ is in this circumstance arranged in *chronological order*. It is important to recognize that, in the most general case, Hamiltonian operators at different times need not commute; it is therefore necessary to adhere to this time sequence.

In the case that

$$[\mathscr{H}(t'), \mathscr{H}(t'')] = 0 \tag{4.8.8}$$

the preceding restriction can be relaxed and the formula for the evolution operator simplified. In a two-dimensional domain of integration it is easily verified that

$$\int_{t_0}^{t} \int_{t_0}^{t'} f(t', t'')\, \mathrm{d}t'\, \mathrm{d}t'' = \tfrac{1}{2} \int_{t_0}^{t} \int_{t_0}^{t} f(t', t'')\, \mathrm{d}t'\, \mathrm{d}t''. \tag{4.8.9}$$

In general,

$$\int_{t_0}^{t} \int_{t_0}^{t'} \cdots \int_{t_0}^{t^{(n-1)}} f(t', t'' \ldots t^{(n)})\, \mathrm{d}t'\, \mathrm{d}t'' \ldots \mathrm{d}t^{(n)}$$
$$= \frac{1}{n!} \int_{t_0}^{t} \int_{t_0}^{t} \cdots \int_{t_0}^{t} f(t', t'' \ldots t^{(n)})\, \mathrm{d}t'\, \mathrm{d}t'' \ldots \mathrm{d}t^{(n)}. \tag{4.8.10}$$

Assuming (4.8.8) and applying (4.8.10) to the expansion (4.8.5)

$$\int_{t_0}^{t} \int_{t_0}^{t'} \cdots \int_{t_0}^{t^{(n-1)}} \mathscr{H}(t')\mathscr{H}(t'') \ldots \mathscr{H}(t^{(n)})\, \mathrm{d}t'\, \mathrm{d}t'' \ldots \mathrm{d}t^{(n)}$$
$$= \frac{1}{n!} \left[\int_{t_0}^{t} \mathscr{H}(t')\, \mathrm{d}t' \right]^n. \tag{4.8.11}$$

Accordingly,

$$\mathscr{U}(t, t_0) = \sum_{n=0}^{\infty} \frac{(i\hbar)^{-n}}{n!} \left[\int_{t_0}^{t} \mathscr{H}(t')\, \mathrm{d}t' \right]^n. \tag{4.8.12}$$

Making use of exponential operator notation

$$\mathrm{e}^{\mathscr{A}} \equiv \sum_{n=0}^{\infty} \mathscr{A}^n / n!, \tag{4.8.13}$$

the evolution operator can be expressed in the compact form:

$$\mathscr{U}(t, t_0) = \exp\left[-\frac{i}{\hbar} \int_{t_0}^{t} \mathscr{H}(t')\, \mathrm{d}t' \right]. \tag{4.8.14}$$

The more general result (4.8.5) can analogously be written

$$\mathscr{U}(t, t_0) = \mathscr{P} \exp\left[-\frac{i}{\hbar} \int_{t_0}^{t} \mathscr{H}(t')\, \mathrm{d}t' \right] \tag{4.8.15}$$

where $\mathscr{P}$ is Dyson's chronological-ordering operator—which has the effect of permuting the n time variables in every term of the expansion (4.8.12) according to the order (4.8.7).

For conservative dynamical systems—Hamiltonian independent of time—the evolution operator simplifies to

$$\mathscr{U}(t, t_0) = \mathscr{U}(t - t_0) = \exp\left[-\frac{i}{\hbar}(t - t_0)\mathscr{H}\right]. \qquad (4.8.16)$$

which depends only on the time difference $t - t_0$. This last result can alternatively be derived by representing the solution to (4.1.1) in a Taylor series expansion

$$\Psi(q, t) = \sum_{n=0}^{\infty} \frac{(t - t_0)^n}{n!}\left(\frac{\partial^n \Psi}{\partial t^n}\right)_{t=t_0}, \qquad (4.8.17)$$

noting that

$$\left(\frac{\partial^n \Psi}{\partial t^n}\right)_{t=t_0} = (i\hbar)^{-1}\mathscr{H}^n\,\Psi(q, t_0), \qquad (4.8.18)$$

and using (4.8.13).

As demonstrated in the preceding calculations, the initial state of a quantum-mechanical system suffices to determine—via the time-dependent Schrödinger equation—both its past and future behaviour. Thus, a quantum system not externally perturbed evolves in an exactly predictable manner. This constitutes, in fact, a quantum analog of the causality or determinacy principle. Quantum causality pertains, however, only to specification of states in terms of the wavefunction $\Psi(q, t)$. In classical mechanics, causality implies that knowledge of the initial positions and velocities (or momenta) of the particles of a system enables complete prediction of its future, as well as its history. The common notion that quantum behaviour is non-causal applies only to these classical configuration variables. Indeterminacy is, in fact, inherent even in the initial specification of a system, measurements on the same quantum state not being, in general, reproducible (cf. Section 3.6).

4.9 Properties of the Evolution Operator

The evolution of a quantum system from time t' to time t can be represented by

$$\Psi(q, t) = \mathscr{U}(t, t')\,\Psi(q, t'), \qquad (4.9.1)$$

in which $\mathscr{U}(t, t')$ takes the appropriate form (4.8.14), (4.8.15) or (4.8.16). For a conservative system, for example,

$$\mathscr{U}(t, t') = \exp\left[-\frac{i}{\hbar}(t - t')\mathscr{H}\right]. \qquad (4.9.2)$$

Equation (4.9.1) represents, in abstract form, the solution of the time-dependent Schrödinger equation with initial condition $\Psi(q, t')$. It is clear from the structure of this operator relation that

$$\mathscr{U}(t, t) = 1 \tag{4.9.3}$$

and that

$$\mathscr{U}(t', t) = \mathscr{U}^{-1}(t, t'). \tag{4.9.4}$$

Evolution operators in successive time intervals possess the group property, i.e.,

$$\mathscr{U}(t, t')\,\mathscr{U}(t', t'') = \mathscr{U}(t, t''). \tag{4.9.5}$$

The hermitian property of the Hamiltonian operator implies the permanence of normalization (cf. 4.3.4):

$$(\psi(q, t), \psi(q, t)) - (\psi(q, t'), \psi(q, t')) = 0. \tag{4.9.6}$$

Substituting (4.9.1), we have

$$\begin{aligned}(\mathscr{U}(t, t')\,\psi(q, t'), \mathscr{U}(t, t')\,\psi(q, t')) - (\psi(q, t'), \psi(q, t'))&\\ = (\psi(q, t'), [\mathscr{U}^{\dagger}\mathscr{U} - 1]\,\psi(q, t')) = 0,&\end{aligned} \tag{4.9.7}$$

showing that $\mathscr{U}(t, t')$ is a unitary operator:

$$\mathscr{U}^{\dagger}(t, t') = \mathscr{U}^{-1}(t, t'). \tag{4.9.8}$$

This property is also implied by the structure of $\mathscr{U}$ (cf. 4.8.15), as an exponential of a skew-hermitian operator $\mathrm{i}\mathscr{H}$. By virtue of (4.9.4) and (4.9.8), hermitian conjugation of $\mathscr{U}$ is equivalent to time reversal:

$$\mathscr{U}^{\dagger}(t, t') = \mathscr{U}(t', t). \tag{4.9.9}$$

Any operator $\mathscr{A}$ which commutes with $\mathscr{H}$ also commutes with $\mathscr{U}$:

$$[\mathscr{A}, \mathscr{H}] = 0 \Rightarrow [\mathscr{A}, \mathscr{U}] = 0. \tag{4.9.10}$$

If, in addition, $\mathscr{A}$ is time-independent, then

$$\begin{aligned}\bar{a}(t) = (\Psi(q, t), \mathscr{A}\Psi(q, t)) &= (\Psi(q, t)\,\mathscr{U}^{\dagger}(t', t)\,\mathscr{A}\,\mathscr{U}(t', t)\,\Psi(q, t'))\\ &= (\Psi(q, t'), \mathscr{A}\Psi(q, t')) = \bar{a}(t'),\end{aligned} \tag{4.9.11}$$

showing quite explicitly that the corresponding dynamical variable is a constant of the motion. The Hamiltonian commutes with $\mathscr{U}$ if (4.8.8) applies, but it is not a constant of the motion unless it is also time-independent. Similar conclusions were arrived at in Section 4.5, on the basis of Heisenberg's equation of motion.

By introducing (4.9.1) into the time-dependent Schrödinger equation, it follows that $\mathscr{U}(t, t')$ is itself a solution:

$$i\hbar \frac{\partial}{\partial t} \mathscr{U}(t, t') = \mathscr{H}(t)\, \mathscr{U}(t, t') \tag{4.9.12}$$

with the initial condition (4.9.3). This operator equation can be solved to give $\mathscr{H}$ in terms of $\mathscr{U}$, making use of (4.9.8):

$$\mathscr{H} = i\hbar \frac{\partial \mathscr{U}}{\partial t} \mathscr{U}^{\dagger}. \tag{4.9.13}$$

From the adjoint of either (4.9.12) or (4.9.13), we have also

$$\mathscr{H} = -i\hbar \mathscr{U} \frac{\partial \mathscr{U}^{\dagger}}{\partial t}. \tag{4.9.14}$$

Some advanced formulations of quantum dynamics make reference to infinitesimal unitary transformations associated with time evolution. For sufficiently small Δt,

$$\Psi(q, t + \Delta t) = \Psi(q, t) + \frac{\partial \Psi}{\partial t} \Delta t = [1 + (i\hbar)^{-1} \Delta t \mathscr{H}]\, \Psi(q, t). \tag{4.9.15}$$

One can thus define an evolution operator for infinitesimal transformations:

$$\mathscr{U}(t + \Delta t, t) = 1 - \frac{i}{\hbar} \Delta t \mathscr{H}, \tag{4.9.16}$$

which applies whatever the form of the Hamiltonian. For a conservative system, we have, taking $\Delta t = (t - t')/n$,

$$\mathscr{U}(t, t') = \lim_{n \to \infty} \left[1 - \frac{i}{\hbar} \frac{(t - t')}{n} \mathscr{H} \right]^n = \exp\left[-\frac{i}{\hbar}(t - t')\mathscr{H} \right] \tag{4.9.17}$$

in agreement with (4.9.2).

4.10 Stationary States

An important category of solutions to the time-dependent Schrödinger equation are those corresponding to *stationary states*, in which the wavefunction is separable in configuration and time variables, i.e.,

$$\Psi(q, t) = \psi(q) f(t). \tag{4.10.1}$$

Stationary states are possible only for conservative dynamical systems, in which the Hamiltonian operator is independent of time. Substituting (4.10.1) into (4.1.1), we obtain

$$i\hbar\, \psi(q) \dot{f}(t) = f(t)\, \mathscr{H} \psi(q), \tag{4.10.2}$$

the Hamiltonian working only on the configuration-space factor. Dividing (4.10.2) by $\psi(q)f(t)$ thus effects a separation of variables:

$$i\hbar\frac{\dot{f}(t)}{f(t)} = \frac{\mathscr{H}\psi(q)}{\psi(q)} = \text{const.} \tag{4.10.3}$$

The second equality evidently corresponds to the *time-independent Schrödinger equation* (cf. 3.7.9)

$$\mathscr{H}\psi(q) = E\psi(q), \tag{4.10.4}$$

the separation constant being identified with the energy of the system. The first equality in (4.10.3) gives a first-order linear differential equation with the solution

$$f(t) = f(0)\exp(-iEt/\hbar). \tag{4.10.5}$$

Wavefunctions for stationary states can accordingly be represented

$$\Psi_n(q,t) = \psi_n(q)\exp(-i\omega_n t) \tag{4.10.6}$$

in which the integration constant $f(0)$ is absorbed into $\psi_n(q)$. For convenience we have introduced the *eigenfrequencies*

$$\omega_n \equiv E_n/\hbar, \tag{4.10.7}$$

with units of angular frequency (radians/sec).

The form of (4.10.6) can be deduced more elegantly by assuming an initial state $\Psi(q,0) = \psi_n(q)$, one of the eigenfunctions of $\mathscr{H}$. Then, applying the evolution operator in the form (4.8.16),

$$\Psi_n(q,t) = \exp\left[-\frac{it}{\hbar}\mathscr{H}\right]\psi_n(q) = \sum_{k=0}^{\infty}\frac{(-it/\hbar)^k}{k!}E_n^k\psi_n(q) = \psi_n(q)\exp(-i\omega_n t). \tag{4.10.8}$$

From still another point of view, (4.10.6) reflects the time-displacement invariance of the Hamiltonian. Since, for a conservative system, $\mathscr{H}(t') = \mathscr{H}(t)$, the eigenfunctions at time t' must differ by no more than a phase factor from those at time t, i.e.

$$\Psi_n(q,t') = e^{i\alpha(t,t')}\Psi_n(q,t). \tag{4.10.9}$$

At time t'',

$$\Psi_n(q,t'') = e^{i\alpha(t,t')}e^{i\alpha(t',t'')}\Psi_n(q,t), \tag{4.10.10}$$

but also

$$\Psi_n(q,t'') = e^{i\alpha(t,t'')}\Psi_n(q,t). \tag{4.10.11}$$

The phase factors must consequently have a linear form such as

$$\alpha(t, t') = \omega_n(t - t'), \tag{4.10.12}$$

the ω_n being constants characteristic of each eigenstate. Finally, by setting $t' = 0$ in (4.10.9), we obtain the form of (4.10.6).

A. *Quasistationary States*

A dynamical system fulfilling condition (4.8.8), in which Hamiltonians at different points in time commute with one another, might be termed *quasi-conservative.* A possible realization might be a Hamiltonian containing a potential-energy term $V(t)$, independent of q or p. As demonstrated in Section 3.7B, commuting operators possess simultaneous eigenfunctions. There must therefore exist a complete set of functions $\psi_n(q)$ satisfying

$$\mathscr{H}(t)\psi_n(q) = E_n(t)\psi_n(q), \tag{4.10.13}$$

with time-dependent eigenvalues. Applying the solution operator in the form (4.8.14) to a presumed initial state $\Psi(q, 0) = \psi_n(q)$, we find

$$\Psi_n(q, t) = \psi_n(q) \exp\left[-\frac{i}{\hbar}\int_0^t E_n(t')\,\mathrm{d}t'\right]. \tag{4.10.14}$$

Time and configuration variables are indeed separable, suggesting the designation *quasistationary*. This result reduces, of course, to (4.10.6) when $\mathscr{H}$ is time-independent.

B. *Cancellation of Time Dependence*

Time-dependence enters into stationary-state wavefunctions only as phase factors of unit magnitude, i.e.,

$$|e^{-i\omega t}| = 1. \tag{4.10.15}$$

The time-independent amplitude $\psi(q)$ can accordingly be used in place of the full wavefunctions $\Psi(q, t)$ in many quantum-mechanical formulas. For example, the probability density (3.2.1) for a stationary state reduces, by virtue of (4.10.15), to

$$\rho(q) = \psi^*(q)\psi(q). \tag{4.10.16}$$

Eigenvalue equations for time-independent operators (cf. Section 3.7) are equally well satisfied by the time-independent functions $\phi_n(q)$. In the formulas for expectation values of time-independent operators—(3.6.9) and following—$\psi(q)$ can likewise be used in place of $\Psi(q, t)$. It is evident, therefore, that the physical properties of stationary states are independent of time—which accounts, of course, for this designation.

The complex time-dependent factors $\exp(-i\omega_n t)$ also cancel in the formulas of stationary perturbation theory. If the time-dependent eigenfunctions $\Psi_n(q, t)$ (4.10.6) are used to evaluate matrix elements, we have

$$V_{mn}(t) = (\Psi_m, \mathscr{V}\Psi_n) = (\psi_m, \mathscr{V}\psi_n)\exp(i\omega_{mn}t) \tag{4.10.17}$$

in terms of the Bohr frequency

$$\omega_{mn} \equiv \omega_m - \omega_n = (E_m - E_n)/\hbar. \tag{4.10.18}$$

In the first and second-order energy formulas,

$$E_m^{(1)} = V_{mm}; \quad E_m^{(2)} = -\mathop{\mathbf{S}'}_n \frac{V_{mn}V_{nm}}{E_n^{(0)} - E_m^{(0)}}, \tag{4.10.19}$$

there is thus cancellation of the time factors, by virtue of (4.10.15). One can therefore use in (4.10.19), as well as other perturbation formulas, time-independent matrix elements defined according to

$$V_{mn} = (\psi_m, \mathscr{V}\psi_n). \tag{4.10.20}$$

4.11 Nonstationary States

Nonstationary states are characterized by wavefunctions $\Psi(q, t)$ with non-separable time dependence. We consider first the case of a conservative system. Any physically-meaningful wavefunction $\Psi(q, t)$ for such a system must conform to the same analytic and boundary conditions which govern its eigenfunctions $\Psi_n(q, t)$. In accordance with Section 3.9, it should therefore be possible to expand an arbitrary nonstationary wavefunction in terms of the eigenfunctions $\psi_n(q)$ of the associated time-independent Schrödinger equation. In particular, for the initial configuration (at $t = 0$)

$$\Psi(q, 0) = \mathop{\mathbf{S}}_n c_n\psi_n(q). \tag{4.11.1}$$

Now, applying the evolution operator in the form (4.8.16), we obtain†

$$\Psi(q, t) = \mathop{\mathbf{S}}_n c_n \exp(-i\omega_n t)\psi_n(q), \tag{4.11.2}$$

or, introducing the time-dependent eigenfunctions,

$$\Psi(q, t) = \mathop{\mathbf{S}}_n c_n\Psi_n(q, t). \tag{4.11.3}$$

† This result also follows by making use of the expansion for $\Psi(q, t)$ at arbitrary t

$$\Psi(q, t) = \mathop{\mathbf{S}}_n c_n(t)\psi_n(q).$$

Substituting into the time-dependent Schrödinger equation, we obtain

$$i\hbar \mathop{\mathbf{S}}_n \dot{c}_n(t)\psi_n(q) = \mathop{\mathbf{S}}_n c_n(t)\mathscr{H}\psi_n(q).$$

It follows that the time-dependent coefficients have the form

$$c_n(t) = c_n \exp(-i\omega_n t).$$

This agrees, in fact, with a well-known result in the theory of partial differential equations, namely that a general solution can be represented as a linear combination of separable (stationary) solutions, with coefficients c_n being determined by the initial conditions. From a physical viewpoint, (4.11.3) is, moreover, a particular instance of the superposition principle (3.9.20).

An interesting analogy can be drawn between the time-development of a nonstationary quantum state, as represented by (4.11.2), and the emission of sound waves by a solid body set into massive vibration. In the acoustical case, the response of the system can be formally represented as a superposition of its normal modes. The object thereby implicitly exhibits its spectrum of natural resonance frequencies, which could be determined by Fourier analysis. The evolving quantum system, is likewise a superposition of stationary modes. It is convenient to define the overlap integral

$$f(t) \equiv (\Psi(q,0), \Psi(q,t)) = \mathop{\mathsf{S}}_{n} |c_n|^2 \exp(-i\omega_n t) \tag{4.11.4}$$

which has the form of a correlation function. The Fourier transform of (4.11.4) gives the *spectral function*

$$g(\omega) = \frac{1}{2\pi}\int_{-\infty}^{\infty} f(t)\, e^{i\omega t}\, dt = \mathop{\mathsf{S}}_{n} |c_n|^2 \delta(\omega - \omega_n), \tag{4.11.5}$$

having used (A.15) to get the deltafunction. In this way, the complete spectrum of eigenfrequencies ω_n is, in principle, exhibited.†

It is interesting to observe that a quantum-mechanical wavefunction which is other than an exact eigenfunction represents, technically speaking, a nonstationary state. Of course, if the deviation from exactness is small, the oscillations about, say Ψ_0, in the hypothetical time evolution of the state will be of relatively small amplitude. Correspondingly the expectation value $\langle \mathscr{H} \rangle$ will approximate E_0 (with $\langle \mathscr{H} \rangle > E_0$, of course, by the variational principle).

For nonconservative systems (excluding the quasiconservative case), there exist no stationary states. The wavefunction can still be represented in a double Fourier expansion of the form

$$\Psi(q,t) = \mathop{\mathsf{S}}_{n} c_n(t)\phi_n(q). \tag{4.11.6}$$

Here $\{\phi_n(q)\}$ is some convenient complete set of basis functions, there being, of course, no eigenfunctions of $\mathscr{H}(t)$.

† I have suggested this as the basis of a computational technique for determining energy eigenvalue spectra of quantum-mechanical systems. See S. M. Blinder, *J. Chem. Phys.* **41**, 3412 (1964); *Int. J. Quantum Chem.* **1**, 271 (1967).

4.12 Metastable States

Such phenomena as radioactive decay and spontaneous emission of radiation involve excited states of relatively long duration, described as being *metastable*. A useful model for such systems utilizes an effective Hamiltonian which is non-hermitian.† In the simplest instance, a constant imaginary term is added such that

$$\mathscr{H}^{\mathrm{eff}} = \mathscr{H} - \frac{i\hbar}{2}\gamma . \tag{4.12.1}$$

The corresponding time-dependent Schrödinger equation is

$$i\hbar\frac{\partial\Psi}{\partial t} = \left(\mathscr{H} - \frac{i\hbar}{2}\gamma\right)\Psi . \tag{4.12.2}$$

By a calculation analogous to that in Section 4.3, the normalization integral varies with time according to

$$\frac{\mathrm{d}}{\mathrm{d}t}(\Psi, \Psi) = -\gamma(\Psi, \Psi), \tag{4.12.3}$$

so that

$$(\Psi, \Psi) = (\Psi, \Psi)_0\, \mathrm{e}^{-\gamma t} \tag{4.12.4}$$

Thus the number of systems in the state Ψ decays exponentially with a time constant γ. This means that the average lifetime of the state is given by

$$\Delta t = 1/\gamma \tag{4.12.5}$$

From another point of view, the continuity equation (cf. Section 4.4) based on (4.12.1) takes the form

$$\frac{\partial\rho}{\partial t} + \nabla\cdot\mathbf{j} = -\gamma\rho \tag{4.12.6}$$

in which the source term (cf. 2.1.9) represents the decay of probability.

If the operator $\mathscr{H}$ is time-independent, eqn (4.12.2) has separable solutions of the form

$$\Psi(q, t) = \psi_n(q)\exp(-i\omega_n t)\exp(-\gamma t/2) \tag{4.12.7}$$

in which ψ_n is an eigenfunction of $\mathscr{H}$ with eigenfrequency ω_n. The wave-function (4.12.7) does not, however, represent a monochromatic energy state. This is shown by evaluating the correlation function (4.11.4):

$$f(t) = \exp(-i\omega_n t)\exp(-\gamma t/2) \qquad (t \geqslant 0) . \tag{4.12.8}$$

† This was first applied to the problem of nuclear α-decay by G. Gamov, *Z. Physik* **51**, 204 (1928). An analog in classical optics is the use of complex index of refraction to represent absorbing media.

Fourier transformation gives the spectral function†

$$g(\omega) = \frac{1}{2\pi}[\gamma/2 - i(\omega - \omega_n)]^{-1}. \tag{4.12.9}$$

Its significance resides in the real part

$$\text{Re}\, g(\omega) = \frac{\gamma/4\pi}{(\omega - \omega_n)^2 + \gamma^2/4}, \tag{4.12.10}$$

which has the form of a Lorentzian distribution (Fig. 4.1) centred at $\omega = \omega_n$. The parameter γ represents the frequency separation between points of half-maximum amplitude. The decaying system represented by (4.12.7) is evidently

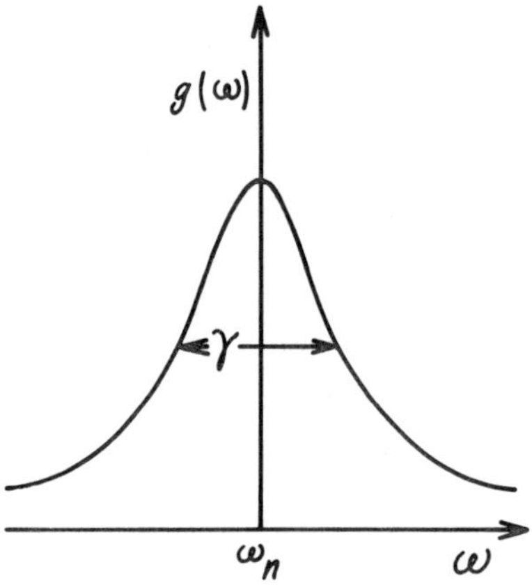

Fig. 4.1. Lorentzian distribution

$$g(\omega) = \frac{\gamma/2\pi}{(\omega - \omega_n)^2 + \gamma^2/4}$$

(normalized to 1).

not in a true energy eigenstate. Results of energy measurements will be in accord with the distribution (4.12.10), characterized by a mean energy uncertainty

$$\Delta E \sim \hbar\gamma. \tag{4.12.11}$$

As a consequence, the spectroscopic transition from this level to the ground state (assumed sharp) will exhibit a natural linewidth of the order of

$$\Delta\omega \sim \gamma \tag{4.12.12}$$

† It is assumed for simplicity that $f(t) = 0$ for $t < 0$. The spectral function is accordingly normalized to $\frac{1}{2}$ rather than 1.

apart from other (usually more drastic) sources of line broadening such as the Doppler effect, collisions and instrumental imperfections. For transitions to other metastable states, both the initial and final energy uncertainties contribute to spectral linewidth.

Metastable states are sometimes described as "weakly quantized". They are intermediate in character between discrete and continuum states. Their spectral resolution shows that they can be built up by superposition of continuum eigenfunctions in such a way as to approximate the behaviour of discrete eigenfunctions. The decay process accordingly represents a transition from a localized to a nonlocalized state of comparable energy. Every excited state is, to some degree, metastable.

4.13 Energy–Time Uncertainty Relation

The 4-vector quantization prescription which formally leads to the time-dependent Schrödinger equation can likewise be applied to obtain a fourth uncertainty relation of the form (3.6.23). This is obviously

$$\Delta x_4 \Delta p_4 \geqslant h/4\pi \tag{4.13.1}$$

or†

$$\Delta t \, \Delta E \geqslant h/4\pi \, . \tag{4.13.2}$$

Physically this means that a measurement of energy carried out within a time interval Δt is subject to an irreducible uncertainty ΔE. Thus a metastable state of average lifetime Δt is associated with an energy band of width $\Delta E \sim \hbar/\Delta t$, in agreement with the results of the preceding section. To determine exactly the energy of a system would require, in concept, an infinitely long time. This is in accord with the essentially time-independent nature of stationary states.

The energy-time uncertainty relation cannot be derived rigorously outside of relativistic quantum mechanics. However, the following physical argument will perhaps make the principle plausible within a nonrelativistic context. Suppose a particle of energy $E = p^2/2m$ is subject to a momentum uncertainty Δp. The associated energy uncertainty is then

$$\Delta E = \frac{p \Delta p}{m} = v \Delta p, \tag{4.13.3}$$

where v is the average classical velocity. If the particle is observed within the time interval Δt, the position uncertainty Δx owing to this velocity will be given by

$$\Delta x = v \Delta t \, . \tag{4.13.4}$$

† The energy–time uncertainty relation was first proposed by N. Bohr, *Nature*, **121**, 580 (1928).

Eliminating v between (4.13.4) and (4.13.3), we find

$$\Delta E\,\Delta t = \Delta p\,\Delta x \geqslant h/4\pi, \tag{4.13.5}$$

in agreement with (4.13.2).

4.14 Heisenberg Picture

The formulation of quantum dynamics employed up to this point is referred to as *Schrödinger picture*. Its characteristic features are (i) wavefunctions $\Psi(q, t)$ which depend on the time and satisfy the time-dependent Schrödinger equation, (ii) operators representing dynamical variables which are independent of time (excepting those cases in which the variable has explicit time dependence). In abstract geometrical terms, Schrödinger picture can be associated with a coordinate system in Hilbert space in which dynamical variables are fixed but state vectors are moving.

Possibilities for alternative formulations exist, however, because both wavefunctions and operators are abstract quantities not directly accessible to measurement. Quantities which *do* have objective physical reality occur as scalars in Hilbert space: eigenvalues, probability densities, expectation values and transition probabilities.

Transformation of coordinates in Hilbert space—perhaps to a moving coordinate system—is effected by means of a unitary transformation. Every wavefunction Ψ and operator $\mathscr{A}$ in the original representation is transformed according to

$$\Psi = \mathscr{U}\Psi' \tag{4.14.1}$$

and

$$\mathscr{A}' = \mathscr{U}^\dagger \mathscr{A}\mathscr{U} \tag{4.14.2}$$

where $\mathscr{U}$ is a unitary operator:

$$\mathscr{U}\mathscr{U}^\dagger = \mathscr{U}^\dagger \mathscr{U} = 1. \tag{4.14.3}$$

Eigenvalue equations are preserved in form under such transformations, since

$$\mathscr{A}\Phi_n = a_n\Phi_n \Rightarrow \mathscr{U}\mathscr{A}'\mathscr{U}^\dagger\,\mathscr{U}\Phi_n' = a_n\mathscr{U}\Phi_n', \tag{4.14.4}$$

so that

$$\mathscr{A}'\Phi_n' = a_n\Phi_n', \tag{4.14.5}$$

with the same eigenvalues a_n. Similarly, every scalar product is invariant under the transformation, for example,

$$(\Phi', \mathscr{A}'\Psi') = (\mathscr{U}^\dagger\,\Phi, \mathscr{U}^\dagger\,\mathscr{A}\mathscr{U}\mathscr{U}^\dagger\,\Psi) = (\Phi, \mathscr{A}\Psi). \tag{4.14.6}$$

Heisenberg picture is established by choosing

$$\mathscr{U} = \mathscr{U}(t, t_0), \tag{4.14.7}$$

the evolution operator from fixed time t_0 to variable time t. Equation (4.9.8) shows that $\mathscr{U}(t, t_0)$ is unitary. Let the superscripts S and H refer to quantities in Schrödinger and Heisenberg pictures, respectively. From (4.14.1), using (4.9.9) and (4.9.1),

$$\Psi^{\mathrm{H}}(q, t) = \mathscr{U}^{\dagger}(t, t_0)\Psi^{\mathrm{S}}(q, t) = \mathscr{U}(t_0, t)\Psi^{\mathrm{S}}(q, t), \tag{4.14.8}$$

so that

$$\Psi^{\mathrm{H}}(q, t) = \Psi^{\mathrm{S}}(q, t_0). \tag{4.14.9}$$

The concommitant operator transformation (4.14.2) gives

$$\mathscr{A}^{\mathrm{H}}(t) = \mathscr{U}^{\dagger}(t, t_0)\mathscr{A}^{\mathrm{S}}\mathscr{U}(t, t_0). \tag{4.14.10}$$

Heisenberg picture corresponds to a coordinate system "rotating" in Hilbert space relative to Schrödinger picture. In this new description, the state vector is brought to rest, retaining simply its initial structure at time t_0. In contrast, dynamical variables are set into motion, their temporal variation being linked with the dynamical evolution of the system. At time $t = t_0$, since $\mathscr{U}(t_0, t_0) = 1$, the two pictures coincide:

$$\Psi^{\mathrm{H}}(q, t_0) = \Psi^{\mathrm{S}}(q, t_0) \tag{4.14.11}$$

and

$$\mathscr{A}^{\mathrm{H}}(t_0) = \mathscr{A}^{\mathrm{S}}. \tag{4.14.12}$$

The relationship between Schrödinger and Heisenberg pictures can be viewed in another way. The expectation value of an observable A is given by

$$\bar{a}(t) = (\Psi^{\mathrm{S}}, \mathscr{A}^{\mathrm{S}}\Psi^{\mathrm{S}}) = (\Psi^{\mathrm{H}}, \mathscr{A}^{\mathrm{H}}\Psi^{\mathrm{H}}), \tag{4.14.13}$$

varying, in general, with time. This time dependence arises, in the Schrödinger picture, because Ψ^{S} changes with time, in the Heisenberg picture, because $\mathscr{A}^{\mathrm{H}}$ changes with time. Equation (4.14.13) shows that state vectors and operators do, nevertheless, preserve their relative orientations in Hilbert space, irrespective of their absolute motions.

The Hamiltonian transforms according to

$$\mathscr{H}^{\mathrm{H}} = \mathscr{U}^{\dagger}\mathscr{H}^{\mathrm{S}}\mathscr{U}. \tag{4.14.14}$$

For conservative (and quasiconservative) systems, when $\mathscr{H}$ commutes with $\mathscr{U}$, then

$$\mathscr{H}^{\mathrm{H}} = \mathscr{H}^{\mathrm{S}} \qquad \text{(conservative case)}. \tag{4.14.15}$$

For nonconservative systems, however, the Hamiltonian can be different in the two pictures. Thus, applying (4.14.14) to (4.9.13) and (4.9.14), we find

$$\mathscr{H}^{\mathrm{H}} = i\hbar\mathscr{U}^{\dagger}\frac{\partial\mathscr{U}}{\partial t} = -i\hbar\frac{\partial\mathscr{U}^{\dagger}}{\partial t}\mathscr{U}. \tag{4.14.16}$$

The fundamental dynamical principle in Schrödinger picture is the time-dependent Schrödinger equation. This has no analog in Heisenberg picture since states are time independent.† Rather, the fundamental equation of motion concerns the rate of change of Heisenberg operators. Taking the time derivative of (4.14.10),

$$\frac{\mathrm{d}\mathscr{A}^{\mathrm{H}}}{\mathrm{d}t} = \frac{\partial\mathscr{U}^{\dagger}}{\partial t}\mathscr{A}^{\mathrm{S}}\mathscr{U} + \mathscr{U}^{\dagger}\frac{\partial\mathscr{A}^{\mathrm{S}}}{\partial t}\mathscr{U} + \mathscr{U}^{\dagger}\mathscr{A}^{\mathrm{S}}\frac{\partial\mathscr{U}}{\partial t}. \tag{4.14.17}$$

Using (4.14.16) and the notation

$$\frac{\partial\mathscr{A}^{\mathrm{H}}}{\partial t} = \mathscr{U}^{\dagger}\frac{\partial\mathscr{A}^{\mathrm{S}}}{\partial t}\mathscr{U}, \tag{4.14.18}$$

we obtain

$$\frac{\mathrm{d}\mathscr{A}^{\mathrm{H}}}{\mathrm{d}t} = \frac{\partial\mathscr{A}^{\mathrm{H}}}{\partial t} + (i\hbar)^{-1}[\mathscr{A}^{\mathrm{H}}, \mathscr{H}^{\mathrm{H}}], \tag{4.14.19}$$

which is Heisenberg's equation of motion. We had previously deduced its analog in Schrödinger picture (cf. 4.5.5, 4.5.7). A fundamental distinction, however, is the complete independence of (4.14.19) from any specification of the state. In Heisenberg picture, $\Psi^{\mathrm{H}}(q, t_0)$ serves as some standard fixed state—merely a "background" to the unfolding dynamics of the system.

An important special case pertains to time-independent dynamical variables which commute with the Hamiltonian. These are *constants of the motion*, already discussed in Section 4.5. The associated operators remain

† Eigenvectors in Heisenberg picture do, however, follow their moving operators. Let Φ^{S} be an eigenvector of some time-independent operator in Schrödinger picture (this is not itself a state function). Then

$$\Phi^{\mathrm{S}}(q) = \mathscr{U}(t, t_0)\Phi^{\mathrm{H}}(q, t).$$

Since Schrödinger eigenvectors are time-independent,

$$\frac{\partial\Phi^{\mathrm{S}}}{\partial t} = \frac{\partial\mathscr{U}}{\partial t}\Phi^{\mathrm{H}} + \mathscr{U}\frac{\partial\Phi^{\mathrm{H}}}{\partial t} = 0.$$

Premultiplying by $\mathscr{U}^{\dagger}$ and using (4.14.16), we obtain the equation of motion for the corresponding Heisenberg eigenvector:

$$i\hbar\frac{\partial\Phi^{\mathrm{H}}}{\partial t} = -\mathscr{H}^{\mathrm{H}}\Phi^{\mathrm{H}}.$$

The occurrence of the minus sign relative to (4.1.1) is associated with "backward rotation" of these eigenvectors in the moving coordinate system.

time independent in Heisenberg picture. In fact, by virtue of (4.9.10), these operators are the same in both pictures:

$$\mathscr{A}^{\mathrm{H}} = \mathscr{A}^{\mathrm{S}} \qquad \text{(constant of the motion).} \tag{4.14.20}$$

The Hamiltonian for a conservative system is itself, of course, a constant of the motion.

A. *Relation to Classical Dynamics*

A close formal correspondence exists between Heisenberg picture and classical dynamics. In both theories, consideration is focused on equations of motion for dynamical variables. The time rate of change of a classical dynamical variable $A(q, p, t)$ is given by the Poisson bracket equation of motion (cf. 1.3.12)

$$\frac{\mathrm{d}A}{\mathrm{d}t} = \frac{\partial A}{\partial t} + \{A, H\}. \tag{4.14.21}$$

The structural analogy between the equations of motion (4.14.21) and (4.14.19) suggests a formal correspondence between classical and quantum dynamics. Specifically, a transition is effected by replacing classical dynamical variables by Heisenberg operators, in accord with the quantization prescription connecting Poisson brackets with commutators (cf.3.4.2):

$$\{A, B\} \rightarrow (i\hbar)^{-1}[\mathscr{A}^{\mathrm{H}}, \mathscr{B}^{\mathrm{H}}]. \tag{4.14.22}$$

This procedure applies, of course, only to observables having classical analogs.

When A and B are themselves generalized coordinates or momenta, the fundamental quantum conditions in Heisenberg picture are obtained, viz.,

$$[q_i^{\mathrm{H}}(t), p_j^{\mathrm{H}}(t)] = i\hbar\delta_{ij} \tag{4.14.23}$$

and

$$[q_i^{\mathrm{H}}(t), q_j^{\mathrm{H}}(t)] = [p_i^{\mathrm{H}}(t), p_j^{\mathrm{H}}(t)] = 0. \tag{4.14.24}$$

These are analogous to (3.4.4) and (3.4.5), respectively, in Schrödinger picture, since, in fact, operator commutation relations are invariant under the transformation (4.14.10). An important stipulation, however, is that the Heisenberg operators be taken at the same time.†

† For a free particle in one dimension, by classical analogy,

$$q^{\mathrm{H}}(t) = q^{\mathrm{H}}(0) + \frac{t}{m}p^{\mathrm{H}}(0)$$

We would then find, for example,

$$[q^{\mathrm{H}}(t), q^{\mathrm{H}}(0)] = \frac{i\hbar t}{m}.$$

As a final point, it is interesting to show the formal correspondence between Hamilton's equations (1.2.12) and Ehrenfest's relations (cf. Section 4.6). The latter are Heisenberg's equations of motion for the generalized coordinates and momenta:

$$\frac{\mathrm{d}q_i^{\mathrm{H}}}{\mathrm{d}t} = (i\hbar)^{-1}[q_i^{\mathrm{H}}, \mathscr{H}^{\mathrm{H}}] \tag{4.14.25}$$

and

$$\frac{\mathrm{d}p_i^{\mathrm{H}}}{\mathrm{d}t} = (i\hbar)^{-1}[p_i^{\mathrm{H}}, \mathscr{H}^{\mathrm{H}}]. \tag{4.14.26}$$

According to the discussion in Section 3.4, each p_i obeys the same commutation relations as the operator $-i\hbar\,\partial/\partial q_i$. Analogously, each q_i is isomorphous with $i\hbar\,\partial/\partial p_i$. Consider, however, a commutator of the following structure acting on an arbitrary function:

$$[\partial/\partial q_i, \mathscr{H}]\Psi = \frac{\partial}{\partial q_i}\mathscr{H}\Psi - \mathscr{H}\frac{\partial \Psi}{\partial q_i} = \frac{\partial \mathscr{H}}{\partial q_i}\Psi. \tag{4.14.27}$$

We can write therefore

$$(i\hbar)^{-1}[q_i^{\mathrm{H}}, \mathscr{H}^{\mathrm{H}}] = \frac{\partial \mathscr{H}^{\mathrm{H}}}{\partial p_i^{\mathrm{H}}} \tag{4.14.28}$$

and, analogously,

$$(i\hbar)^{-1}[p_i^{\mathrm{H}}, \mathscr{H}^{\mathrm{H}}] = -\frac{\partial \mathscr{H}^{\mathrm{H}}}{\partial q_i^{\mathrm{H}}}. \tag{4.14.29}$$

Substituting these into (4.14.25) and (4.14.26), respectively, we obtain the operator analogs of Hamilton's equations.

4.15 Matrix Mechanics

When dynamical variables in Heisenberg picture are given matrix representations, in accordance with the considerations of Section 3.11, the resultant formalism is identical with matrix mechanics. The latter was, of course, the earliest formulation of quantum mechanics, developed by Heisenberg, Born and Jordan in 1925. The fundamental postulates of matrix mechanics can be stated as follows.

1. Every dynamical variable A is represented by a hermitian matrix $\mathbb{A}$ (cf. 3.11.12), whose elements are, in general, functions of time.

2. The possible results of individual measurement are the eigenvalues of $\mathbb{A}$.

3. Dynamical variables $A(q, p, t)$ having classical analogs are represented

by matrices $\mathbb{A}(\mathbb{Q}, \mathbb{P}, t)$ containing the same functional dependence on $\mathbb{Q}_i$ and $\mathbb{P}_i$ matrices. The latter are constructed so to satisfy the fundamental matrix commutation relations.

$$[\mathbb{Q}_i, \mathbb{P}_j] \equiv \mathbb{Q}_i \mathbb{P}_j - \mathbb{P}_j \mathbb{Q}_i = i\hbar\delta_{ij}\mathbb{1} \tag{4.15.1}$$

where $\mathbb{1}$ is the unit matrix. The $\mathbb{Q}_i$ and $\mathbb{P}_j$ matrices commute among themselves.

4. The time-dependence of $\mathbb{A}$ is determined by Heisenberg's equation of motion:

$$\frac{\mathrm{d}\mathbb{A}}{\mathrm{d}t} = \frac{\partial \mathbb{A}}{\partial t} + (\mathrm{i}\hbar)^{-1}[\mathbb{A}, \mathbb{H}] \tag{4.15.2}$$

where $\mathbb{H}$ is the Hamiltonian matrix.

Note that matrix mechanics makes no explicit reference to wavefunctions or state vectors. All physically meaningful quantities are represented in terms of matrix elements.

The formalism of matrix mechanics and wave mechanics become equivalent† when the appropriate connection is made between matrix elements and operator integrals. Fundamentally, this is contained in the relation (cf. 3.9.28).

$$A_{mn} = (\Phi_m, \mathscr{A}\Phi_n) \tag{4.15.3}$$

with reference to some appropriate basis $\{\Phi_n\}$.

Heisenberg picture, and its realization in matrix mechanics, is a formalism of distinctive elegance and appealing analogy to classical theory. However, wave mechanics in Schrödinger picture is generally more convenient on the computational level.

† The equivalence of the two theories was established by E. Schrödinger, *Ann. Phys.* **79**, 734 (1926); also C. Eckart, *Phys. Rev.* **28**, 711 (1926).

5

The Free Particle

Many of the equations of quantum mechanics assume their simplest form when applied to the free particle. With mathematical complications thus minimized, the underlying concepts of the quantum theory can be given their most explicit representation. In this chapter, we shall consider the one-dimensional and three-dimensional free particle, something of the nature of continuous spectra and, as applications, the time–evolution of a wavepacket and the elementary formulation of scattering theory.

5.1 One Dimension

For the free particle in one dimension the Schrödinger equation has the form

$$i\hbar\frac{\partial\Psi}{\partial t} = -\frac{\hbar^2}{2m}\frac{\partial^2\Psi}{\partial x^2}. \tag{5.1.1}$$

The stationary solutions are

$$\Psi_k(x, t) = C\,\mathrm{e}^{i(kx-\omega t)} \tag{5.1.2}$$

where the wavenumber k has an unrestricted real spectrum:

$$-\infty < k < \infty \tag{5.1.3}$$

and

$$\omega = \frac{\hbar k^2}{2m}. \tag{5.1.4}$$

The time-independent amplitudes

$$\psi_k(x) = C\,\mathrm{e}^{ikx} \tag{5.1.5}$$

are simultaneous eigenfunctions of the time-independent Schrödinger equation

$$-\frac{\hbar^2}{2m}\frac{\mathrm{d}^2\psi}{\mathrm{d}x^2} = E\psi \tag{5.1.6}$$

and the linear-momentum eigenvalue equation

$$-i\hbar\frac{d\psi}{dx} = p\psi. \tag{5.1.7}$$

The eigenvalues are, respectively,

$$E = \hbar^2k^2/2m \tag{5.1.8}$$

and

$$p = \hbar k \tag{5.1.9}$$

being related by

$$E = p^2/2m \tag{5.1.10}$$

as in classical mechanics. By virtue of (5.1.3), these eigenvalue spectra are continuous, with

$$E \geqslant 0, \qquad -\infty < p < \infty. \tag{5.1.11}$$

These continua arise, of course, from the absence of restrictive boundary conditions. The Schrödinger equation has the alternative solutions:

$$\psi_k(x) = C\begin{cases}\sin kx\\ \cos kx\end{cases} \tag{5.1.12}$$

which are not, however, eigenfunctions of momentum. The alternative forms of the eigenfunctions, (5.1.5) and (5.1.12), have the structure of travelling waves and standing waves, respectively. This is clearly demonstrated by calculation of the probability current, using the one-dimensional form of (4.4.5):

$$j = \frac{\hbar}{2im}\left(\Psi^*\frac{\partial\Psi}{\partial x} - \Psi\frac{\partial\Psi^*}{\partial x}\right). \tag{5.1.13}$$

With (5.1.12), $j = 0$, showing no net flux. With (5.1.5), on the other hand,

$$j = |C|^2\frac{\hbar k}{m} = |C|^2 v_p \tag{5.1.14}$$

representing a probability flux proportional to the classical particle velocity v_p.

5.2 Normalization

Were the eigenfunctions (5.1.5) members of a discrete spectrum, one would fix the constant C so as to fulfil the normalization condition

$$\int_{-\infty}^{\infty} \psi_k^*(x)\psi_k(x)\,dx = 1. \tag{5.2.1}$$

This integral diverges, however, since

$$\int_{-\infty}^{\infty} |e^{ikx}|\,dx = \infty, \tag{5.2.2}$$

demonstrating, in fact, a characteristic behaviour of continuum eigenfunctions. This result is reasonable from a physical point of view, since a momentum eigenstate implies $\Delta p = 0$ in the uncertainty product (3.6.23). We must therefore have $\Delta x = \infty$, representing a state completely delocalized over the infinite domain $-\infty < x < \infty$.

The wave intensity associated with (5.1.2) or (5.1.5) is

$$\rho(x, t) = \Psi_k^*(x, t)\,\Psi_k(x, t) = |C|^2 \tag{5.2.3}$$

a constant for all x and t. Born's interpretation of $\Psi^*\Psi$ as a single-particle probability density (cf. Section 3.2) does not evidently apply in this instance, for otherwise Ψ should be normalizable to unity. One can, however, reinterpret the wave intensity as a measure of *particle density* (in one dimension, particles per unit length). This reduces, in fact, to the Born interpretation for a normalizable one-particle system. In accordance with the suggested generalization, a free-particle eigenstate must represent a hypothetical, infinitely-long beam of monoenergetic particles. The failure of normalization means simply that the beam contains an infinite number of particles. Still, according to (5.2.3), the linear density is finite. And by choosing

$$C = N^{\frac{1}{2}}, \tag{5.2.4}$$

we specify a beam containing N particles per unit length. The probability current (5.1.14) can correspondingly be written

$$j = Nv_p \tag{5.2.5}$$

which is the one-dimensional analog of (4.4.7), representing the number of particles passing a given point per unit time.

Since $\rho(x, t)$ is now interpreted as a particle density (particles/length), a one-dimensional continuum wavefunction must have the dimensions of (particles/length)$^{\frac{1}{2}}$.

It is formally advantageous, according to Sections 3.8 and 3.9, to establish *deltafunction normalization* for the free-particle eigenstates. The overlap integral between two different wavefunctions (5.1.5) gives

$$\int_{-\infty}^{\infty} \psi_k^*(x)\psi_{k'}(x)\,dx = N\int_{-\infty}^{\infty} e^{i(k'-k)x}\,dx = 2\pi N\delta(k - k') \tag{5.2.6}$$

in accordance with the representation (A.15) of the deltafunction. Setting

$N = (2\pi)^{-1}$, we obtain deltafunction-orthonormalized free-particle eigenfunctions†, viz.,

$$\psi_k(x) = (2\pi)^{-\frac{1}{2}} e^{ikx}. \tag{5.2.7}$$

Also in common use is *box normalization. Periodic boundary conditions* are imposed on the wavefunction, whereby

$$\psi(x + L) = \psi(x). \tag{5.2.8}$$

This restricts k to the discrete set of values

$$k = 2n\pi/L, \qquad n = 0, \pm 1, \pm 2 \ldots \tag{5.2.9}$$

which however approaches a continuum as $L \to \infty$. The eigenfunctions

$$\psi_k(x) = L^{-\frac{1}{2}} e^{ikx} \tag{5.2.10}$$

are then orthonormalized in the usual sense within each periodic "box", i.e.,

$$\int_{-L/2}^{L/2} \psi_k^*(x)\, \psi_{k'}(x)\, dx = \delta_{kk'} \tag{5.2.11}$$

Dependence of L will cancel out in all physically meaningful results.

By the symmetry between x and k in these eigenfunctions, the closure relation (3.9.25) is easily demonstrated:

$$\int_{-\infty}^{\infty} \psi_k^*(x)\, \psi_k(x')\, dk = \delta(x - x'). \tag{5.2.12}$$

5.3 Wavepackets

By superposition of the infinite monochromatic waves representing momentum eigenstates, localized *wavepackets* can be constructed. Localization is effected, in concept, by selective interference among matter waves of different wavenumber. Wavepackets correspond more closely to the behaviour of classical particles, exhibiting, for example, particle-like trajectories.

The initial configuration $\Psi(x, 0)$ of a wavepacket can be represented by

$$\Psi(x, 0) = \int_{-\infty}^{\infty} \phi(k)\, \psi_k(x)\, dk \tag{5.3.1}$$

in terms of the eigenfunctions (5.2.7). Specifically, this has the form of a Fourier integral:

$$\Psi(x, 0) = \frac{1}{\sqrt{(2\pi)}} \int_{-\infty}^{\infty} \phi(k)\, e^{ikx}\, dk. \tag{5.3.2}$$

† Momentum is sometimes used as a quantum label. The free-particle eigenfunctions are taken as $\psi_p(x) = h^{-\frac{1}{2}} e^{ipx/\hbar}$, orthonormalized such that $\int_{-\infty}^{\infty} \psi_p^*(x)\psi_{p'}(x)\, dx = \delta(p - p')$.

This is a concrete example of expansion (4.11.1), for a nonstationary state of a conservative system. At a later time t, the wavepacket evolves to (cf. 4.11.2)

$$\Psi(x, t) = \frac{1}{\sqrt{(2\pi)}} \int_{-\infty}^{\infty} \Phi(k) \exp i[kx - \omega(k)t] \, \mathrm{d}k \tag{5.3.3}$$

where

$$\omega(k) = \hbar k^2/2m. \tag{5.3.4}$$

A normalized wavepacket

$$\int_{-\infty}^{\infty} |\Psi(x, t)|^2 \, \mathrm{d}x = 1 \tag{5.3.5}$$

represents a single particle, rather than an infinite beam. By virtue of (4.3.4), normalization persists in time.

5.4 Phase and Group Velocities

The free-particle eigenfunctions

$$\Psi_k(x, t) = \frac{1}{\sqrt{(2\pi)}} \mathrm{e}^{i(kx - \omega t)} \tag{5.4.1}$$

have the general structure of travelling waves. It is instructive therefore to examine the quantum analogs of the various parameters of classical wave theory. For $k > 0$ [$k < 0$], (5.4.1) corresponds to propagation in the positive [negative] x-direction. This is indeed a representation of de Broglie's *matter waves*, the associated wavelength being given by

$$\lambda = \frac{2\pi}{|k|} = \frac{h}{|p|}. \tag{5.4.2}$$

The *phase* of a monochromatic wave is the argument of the exponential function:

$$\phi = kx - \omega t = 2\pi\left(\frac{x}{\lambda} - vt\right). \tag{5.4.3}$$

Loci of constant phase in the moving wave satisfy the condition

$$\mathrm{d}\phi = k\,\mathrm{d}x - \omega\,\mathrm{d}t = 0. \tag{5.4.4}$$

The *phase velocity* (or *wave velocity*) is the speed at which points of constant phase—such as wave crests or troughs—propagate. From (5.4.4), we have

$$v_\phi = \left(\frac{\mathrm{d}x}{\mathrm{d}t}\right)_\phi = \frac{\omega}{k}. \tag{5.4.5}$$

We thus obtain the well-known relationship between wavelength and frequency

$$v_\phi = \lambda \nu \qquad (5.4.6)$$

which is generally valid for all types of wave motion.

For matter waves, the phase can also be expressed

$$\phi = \frac{1}{\hbar}(px - Et) \qquad (5.4.7)$$

so that

$$v_\phi = E/p. \qquad (5.4.8)$$

Specifically,

$$v_\phi = \frac{\hbar k}{2m} = \tfrac{1}{2} v_p \qquad (5.4.9)$$

where v_p is the classical particle velocity.

This result is actually an artifice of the nonrelativistic treatment. Based on the relativistic expressions for energy and momentum:

$$E = mc^2(1 - \beta^2)^{-\frac{1}{2}} \qquad (5.4.10)$$

and

$$p = mv_p(1 - \beta^2)^{-\frac{1}{2}} \qquad (5.4.11)$$

we find

$$v_\phi = c^2/v_p \qquad (5.4.12)$$

or

$$v_\phi v_p = c^2. \qquad (5.4.13)$$

Particle velocities must be less than the speed of light. We arrive thereby at the remarkable result that the phase velocity of relativistic matter waves is *greater* than the speed of light.

Equation (5.4.9) (or its relativistic counterpart) shows that v_ϕ for matter waves depends on the wavenumber k. By contrast, the velocity of light *in vacuo* is a constant independent of frequency. But in material media, the velocity of light *does* depend on its frequency, a phenomenon known as *dispersion.*

Matter waves are rather strongly dispersive. In a wavepacket, each frequency component has a different phase velocity. Thus phase relationships among components will vary to produce a continuous change in the shape of the packet. The probability density associated with the wavepacket (5.3.2) at $t = 0$ and at $t = t$ are, respectively,

$$\rho(x, 0) = \frac{1}{2\pi} \int_{-\infty}^{\infty} \int_{-\infty}^{\infty} \phi(k)\phi^*(k')\, e^{i(k-k')x}\, dk\, dk' \tag{5.4.14}$$

and

$$\rho(x, t) = \frac{1}{2\pi} \int_{-\infty}^{\infty} \int_{-\infty}^{\infty} \phi(k)\phi^*(k') \exp[i(k-k')x] \exp\{-i[\omega(k)-\omega(k')]t\}\, dk\, dk' \tag{5.4.15}$$

If the wavepacket is made up of a group of waves varying over a relatively narrow range of k, it is valid to approximate

$$\omega(k) - \omega(k') \approx \frac{d\omega}{dk}(k - k'). \tag{5.4.16}$$

With (5.4.16) in (5.4.15), the exponential factor can be written

$$\exp\left[i(k - k')\left(x - \frac{d\omega}{dk}t\right)\right]. \tag{5.4.17}$$

Accordingly,

$$\rho(x, t) \approx \rho\left(x - \frac{d\omega}{dk}t, 0\right). \tag{5.4.18}$$

This shows that the wavepacket translates at a velocity $d\omega/dk$ while undergoing no significant distortion in its initial shape. It is suggestive thereby to define a *group velocity*

$$v_g = d\omega/dk. \tag{5.4.19}$$

Alternatively,

$$v_g = dE/dp \tag{5.4.20}$$

and

$$v_g = \frac{d\nu}{d\lambda^{-1}} \tag{5.4.21}$$

The last expression is equivalent to

$$v_g = v_\phi - \lambda \frac{dv_\phi}{d\lambda}. \tag{5.4.22}$$

The group velocity can alternatively be derived by the following argument. At the point x_{max}, corresponding to the maximum value of a wavepacket, there must be maximum reenforcement among the component waves. This means that the maximum number of waves must be in phase with one another.

In mathematical terms, the phase

$$\phi_{max} = kx_{max} - \omega t \tag{5.4.23}$$

must be *stationary* w.r.t. small variations in k:

$$\delta\phi_{max} = (\delta k)x_{max} - (\delta\omega)t = 0. \tag{5.4.24}$$

Noting that

$$\delta\omega = \frac{d\omega}{dk}\,\delta k \tag{5.4.25}$$

we find the velocity of x_{max} given by

$$\frac{dx_{max}}{dt} = \frac{d\omega}{dk} \tag{5.4.26}$$

in agreement with (5.4.19).

In all wave phenomena, the actually observed speed of a pulse is the group velocity. For example, waves in deep water are subject to the approximate dispersion condition $v_\phi \propto \lambda^{\frac{1}{2}}$. Therefore, $v_g \approx \frac{1}{2}v_\phi$: a finite disturbance, such as a boat wake, thus moves at *half* the speed of its component waves. Only in the case of no dispersion (such as light *in vacuo*), does $v_g = v_\phi$, so that pulses travel at the phase velocity.

For matter waves, we find from (5.4.19) and (5.3.4) that

$$v_g = \hbar k/m = v_p \tag{5.4.27}$$

Indeed the group velocity coincides with the classical particle velocity,† quite consistent with the pseudoclassical behaviour of wavepackets.

5.5 Minimum Uncertainty Packet

For a wavepacket $\Psi(x, 0)$ of the form (5.3.2), let the average values of position and momentum be denoted

$$\langle x \rangle \equiv x_0 \tag{5.5.1}$$

† The relativistic group velocity is likewise equal to the particle velocity. Differentiating

$$E^2 = p^2c^2 + m^2c^4$$

we obtain

$$2E\, dE/dp = 2pc^2$$

so that

$$v_g = dE/dp = pc^2/E = v_p$$

and

$$\langle p \rangle \equiv \hbar k_0 . \tag{5.5.2}$$

The corresponding uncertainties—root mean square deviations from the mean—are then given by

$$(\Delta x)^2 = \int_{-\infty}^{\infty} \Psi^*(x, 0)(x - x_0)^2 \Psi(x, 0)\, \mathrm{d}x \tag{5.5.3}$$

and

$$(\Delta p)^2 = \hbar^2 \int_{-\infty}^{\infty} \Psi^*(x, 0)\left(i\frac{\mathrm{d}}{\mathrm{d}x} + k_0\right)^2 \Psi(x, 0)\, \mathrm{d}x. \tag{5.5.4}$$

The momentum expectation value and uncertainty can alternatively be expressed in terms of the Fourier transform:

$$\langle p \rangle = \hbar \int_{-\infty}^{\infty} |\phi(k)|^2 k\, \mathrm{d}k \tag{5.5.5}$$

$$(\Delta p)^2 = \hbar^2 \int_{-\infty}^{\infty} |\phi(k)|^2 (k - k_0)^2\, \mathrm{d}k . \tag{5.5.6}$$

The uncertainties must conform to the inequality

$$\Delta x \Delta p \geqslant h/4\pi \tag{5.5.7}$$

in accordance with the Heisenberg uncertainty principle (3.6.23). It is of interest to deduce the explicit form of the wavepacket for which the minimum value of the uncertainty product applies.

Referring to the derivation in Section 3.6A, it is clear that the limiting form of Schwartz' inequality (3.6.16) applies when f and g are proportional, say,

$$f(x) = \alpha g(x). \tag{5.5.8}$$

In the present application, let

$$f(x) = (x - x_0)\Psi(x, 0) \tag{5.5.9}$$

and

$$g(x) = \hbar\left(\frac{\mathrm{d}}{\mathrm{d}x} - ik_0\right)\Psi(x, 0) . \tag{5.5.10}$$

We obtain thereby the differential equation

$$\alpha\hbar \frac{\mathrm{d}\Psi}{\mathrm{d}x} = [(x - x_0) + i\alpha\hbar k_0]\Psi. \tag{5.5.11}$$

Separating variables and integrating:

$$\alpha\hbar \int_{\Psi(x_0,0)}^{\Psi(x,0)} \mathrm{d}\ln\Psi(x,0) = \int_{x_0}^{x} [(x - x_0) + i\alpha\hbar k_0]\,\mathrm{d}x \tag{5.5.12}$$

which results in

$$\Psi(x,0) = \Psi(x_0,0)\exp\left[\frac{(x - x_0)^2}{2\alpha\hbar} + ik_0(x - x_0)\right]. \tag{5.5.13}$$

For this function to be finite everywhere, α must be a negative constant. Now, let

$$\sigma_0^2 \equiv -\alpha\hbar/2 \tag{5.5.14}$$

and adjust $\Psi(x_0, 0)$ so as to normalize the wavefunction. The result is

$$\Psi(x,0) = (2\pi\sigma_0^2)^{-\frac{1}{4}}\exp\left[-\frac{(x - x_0)^2}{4\sigma_0^2} + ik_0(x - x_0)\right] \tag{5.5.15}$$

which is evidently the quantum-mechanical analog of a particle instantaneously located at $x = x_0$, moving with momentum $p = \hbar k_0$. The corresponding probability density

$$\rho(x,0) = (2\pi\sigma_0^2)^{-\frac{1}{2}}\,\mathrm{e}^{-(x-x_0)^2/2\sigma_0^2} \tag{5.5.16}$$

has the form of a normalized gaussian distribution function. It is easily verified that

$$\Delta x = \sigma_0 \tag{5.5.17}$$

(equal to the gaussian standard deviation) while

$$\Delta p = \hbar/2\sigma_0. \tag{5.5.18}$$

Thus the uncertainty product assumes its minimum value:

$$\Delta x\Delta p = h/4\pi. \tag{5.5.19}$$

5.6 Evolution of a Gaussian Wavepacket

The initial form (5.5.15) of the minimum-uncertainty packet is independent of the dynamical structure of the system. Its subsequent development is, however, determined by the Hamiltonian. This will be worked out for the free particle.

The computation is based on eqns (5.32) and (5.3.3). The first step is to Fourier analyze the initial wavepacket. By the Fourier inversion theorem (cf. B.10)

$$\phi(k) = \frac{1}{\sqrt{(2\pi)}}\int_{-\infty}^{\infty} \Psi(x,0)\,\mathrm{e}^{-ikx}\,\mathrm{d}x. \tag{5.6.1}$$

Substituting (5.5.15), we obtain

$$\phi(k) = (2\pi\sigma_0^2)^{-\frac{1}{4}}\, e^{-ikx_0} \frac{1}{\sqrt{(2\pi)}} \int_{-\infty}^{\infty} \exp\left[-\frac{(x - x_0)^2}{4\sigma_0^2} - i(k - k_0)(x - x_0) \right] dx \tag{5.6.2}$$

In the appropriate variables, the integral represents the Fourier transform of a gaussian (cf. B.13). Accordingly

$$\phi(k) = \left(\frac{2\sigma_0^2}{\pi}\right)^{\frac{1}{4}} e^{-ikx_0}\, e^{-\sigma_0^2(k-k_0)^2}. \tag{5.6.3}$$

We proceed now to resynthesize the wavepacket at time t, in accordance with (5.3.3) and (5.3.4):

$$\Psi(x, t) = \frac{1}{\sqrt{(2\pi)}} \int_{-\infty}^{\infty} \phi(k) \exp[i(kx - \hbar k^2 t/2m)]\, dk. \tag{5.6.4}$$

Substituting (5.6.3),

$$\Psi(x, t) = \left(\frac{2\sigma_0^2}{\pi}\right)^{\frac{1}{4}} e^{-\sigma_0^2 k_0^2} \frac{1}{\sqrt{(2\pi)}} \int_{-\infty}^{\infty} \exp[-\sigma_0 \Sigma k^2 + i(x - x_0 - 2i\sigma_0^2 k_0)k]\, dk \tag{5.6.5}$$

having introduced the complex parameter

$$\Sigma(t) = \sigma_0 \left(1 + \frac{i\hbar t}{2m\sigma_0^2}\right). \tag{5.6.6}$$

The last integral has the form of the inverse transform to the one used above. The result is†

$$\Psi(x, t) = (2\pi\Sigma^2)^{-\frac{1}{4}}\, e^{-\sigma_0^2 k_0^2} \exp\left[-\frac{(x - x_0 - 2i\sigma_0^2 k_0)^2}{4\sigma_0 \Sigma} \right]. \tag{5.6.7}$$

It can be verified by direct substitution that this is indeed a solution of the free-particle time-dependent Schrödinger equation (5.1.1). Note that this wavefunction exhibits time-inversion symmetry (cf. 4.7.4) since

$$\Psi(x, -t) = \Psi^*(x, t). \tag{5.6.8}$$

† This will be rederived in Section 6.5 by use of Green's functions. An alternative approach based on the evolution operator is given in S. M. Blinder, *Am. J. Phys.* **36**, 525 (1958). It is interesting to compare (5.6.7) with a well-known result in classical diffusion theory. The Schrödinger equation for a free particle has the same form as the Fokker–Planck equation for diffusion:

$$\partial\Phi/\partial t = D\nabla^2\Phi$$

where Φ represents the concentration and D, the diffusion coefficient. For diffusion in one dimension from an initial point source of concentration $\Phi(x, 0) = \delta(x)$, the Fokker–Planck equation has the solution

$$\Phi(x, t) = (4\pi Dt)^{-\frac{1}{2}}\, e^{-x^2/4Dt}.$$

After some complex algebra, the probability distribution associated with the wavepacket is shown to be

$$\rho(x, t) = (2\pi\sigma^2)^{-\frac{1}{2}} \exp\left[-\frac{(x - x_0 - \hbar k_0 t/m)^2}{2\sigma^2} \right], \tag{5.6.9}$$

in terms of

$$\sigma(t) \equiv |\Sigma(t)| = \sigma_0 \left(1 + \frac{\hbar^2 t^2}{4m^2\sigma_0^4} \right)^{\frac{1}{2}}. \tag{5.6.10}$$

Remarkably, the wavepacket maintains the gaussian structure (5.5.16) with x_0 generalized to $x_0 + \hbar k_0 t/m$, and σ_0 to $\sigma(t)$. The packet's centre of symmetry moves according to

$$\langle x \rangle = x_0 + \frac{\hbar k_0}{m} t, \tag{5.6.11}$$

thus following the classical particle trajectory. Since this is also the point of maximum wave amplitude, earlier conclusions on group velocity (cf. Section 5.4) are substantiated. The wavepacket halfwidth is given by

$$\Delta x = \sigma(t). \tag{5.6.12}$$

This increases with time in accordance with (5.6.10), thus exhibiting the dispersive character of matter waves.

The momentum expectation value and uncertainty are most easily calculated using (5.6.3) in (5.5.5) and (5.5.6). The results are

$$\langle p \rangle = \hbar k_0 \tag{5.6.13}$$

and

$$\Delta p = \hbar/2\sigma_0. \tag{5.6.14}$$

These agree with (5.5.2) and (5.5.18), respectively, consistent with the fact that the original superposition of momentum eigenstates remains invariant in time. The uncertainty product at time t is given by

$$\Delta x \Delta p = \frac{\hbar}{2} \left(1 + \frac{\hbar^2 t^2}{4m^2\sigma_0^4} \right)^{\frac{1}{2}}. \tag{5.6.15}$$

This is seen to increase monotonically with time from its original minimum value. The more localized is the wavepacket initially, the larger will be its momentum uncertainty and the more rapidly will it diffuse.

Ehrenfest's relations (4.6.7) and (4.6.8) are easily verified for the evolving wavepacket. Thus

$$\frac{d\langle x \rangle}{dt} = \frac{\hbar k_0}{m} = \frac{\langle p \rangle}{m} \tag{5.6.16}$$

showing that the particle is in uniform linear motion (or at rest if $k_0 = 0$), and

$$\frac{d\langle p\rangle}{dt} = 0 \tag{5.6.17}$$

since there are no external forces on a free particle. Moreover,

$$\frac{d^2\langle x\rangle}{dt^2} = 0, \tag{5.6.18}$$

which is the quantum analog of Newton's first law.

Although these analogs of the laws of classical mechanics are rigorously valid, they do not provide a particularly useful description of a particle's motion unless the wavepacket holds together. This applies during a time interval such that

$$\frac{\hbar t}{2m\sigma_0^2} \ll 1. \tag{5.6.19}$$

But by the time

$$t_q \sim m\sigma_0^2/\hbar \tag{5.6.20}$$

the packet has diffused sufficiently that a quantum description becomes mandatory. To illustrate, for an electron initially localized within a radius of 10^{-8} cm, $t_q \sim 10^{-16}$ sec. A molecule 1 Å long has $t_q \sim 10^{-11}$ sec, typical of the time between molecular collisions in a liquid. At the opposite extreme, a 1 g mass of dimension 1 cm would require $\sim 10^{20}$ years to deviate significantly from classical behaviour (the Universe is $\sim 10^{10}$ years old!).

5.7 Three Dimensions : Plane Wave Solutions

The free-particle Schrödinger equation in three dimensions is

$$i\hbar\partial\Psi(\mathbf{r}, t)/\partial t = -\frac{\hbar^2}{2m}\nabla^2\Psi(\mathbf{r}, t). \tag{5.7.1}$$

For stationary solutions,

$$(\nabla^2 + k^2)\Psi(\mathbf{r}) = 0, \tag{5.7.2}$$

where

$$k^2 = 2mE/\hbar^2, \tag{5.7.3}$$

which has the form of the Helmholtz equation. The time-independent Schrödinger equation (5.7.2) is separable in cartesian coordinates (among others). Each factor can be given the structure of (5.2.7), for deltafunction-normalized travelling-wave solutions. Thus

$$\psi_{k_1 k_2 k_3}(xyz) = (2\pi)^{-\frac{3}{2}}\, e^{ik_1 x}\, e^{ik_2 y}\, e^{ik_3 z}, \tag{5.7.4}$$

which can be condensed to†

$$\psi_{\mathbf{k}}(\mathbf{r}) = (2\pi)^{-\frac{3}{2}}\, e^{i\mathbf{k}\cdot\mathbf{r}} \tag{5.7.5}$$

having introduced the *wavevector* (*propagation vector*) **k**. The normalization constant implies, in view of (5.2.4), a standard density of $(2\pi)^{-3}$ particles per unit volume. The orthonormalization condition on these eigenfunctions can be written

$$\int \psi^*_{\mathbf{k}}(\mathbf{r})\psi_{\mathbf{k}'}(\mathbf{r})\, d\tau = \delta(\mathbf{k} - \mathbf{k}'). \tag{5.7.6}$$

Each component $k_1 k_2 k_3$ of the wavevector can take on arbitrary real values, as in (5.1.3). The energy eigenvalues corresponding to (5.7.5) are accordingly

$$E = \frac{\hbar^2}{2m}(k_1^2 + k_2^2 + k_3^2) \tag{5.7.7}$$

which is equivalent to (5.1.8) if k is interpreted as the absolute value of **k**. The corresponding momentum vector eigenvalues are

$$\mathbf{p} = \hbar\mathbf{k}. \tag{5.7.8}$$

The full time-dependent energy-momentum eigenfunctions have the form

$$\Psi_k(\mathbf{r}, t) = (2\pi)^{-\frac{3}{2}}\, e^{i(\mathbf{k}\cdot\mathbf{r} - \omega t)}. \tag{5.7.9}$$

For fixed **k**, the equation

$$\mathbf{k}\cdot\mathbf{r} = \text{const} \tag{5.7.10}$$

determines a plane normal to **k**. The loci of constant phase

$$\phi = \mathbf{k}\cdot\mathbf{r} - \omega t \tag{5.7.11}$$

correspond therefore to an infinite succession of equidistant planes $2\pi/k$ apart propagating in the direction of **k** with a (nonrelativistic) phase velocity

$$\mathbf{v}_\phi = \hbar\mathbf{k}/2m. \tag{5.7.12}$$

Thus (5.7.9) describes solutions to (5.7.1) having the form of *plane waves.*

5.8 Density of States

For the free particle in one dimension, all eigenfunctions with $k \neq 0$ are

† By an argument analogous to that in eqns (4.10.9)–(4.10.12), the structure of these eigenfunctions is determined, in principle, by the translational symmetry of the free-particle Hamiltonian, $\mathscr{H}(\mathbf{r}') = \mathscr{H}(\mathbf{r})$.

doubly degenerate. This is consistent with the equivalence of momentum values p and $-p$ to the same kinetic energy.

In three dimensions, the degeneracy is a more complicated function of energy. Clearly, however, the higher the energy, the more ways there are of composing a vector **k** of given magnitude. Degeneracy will therefore increase monotonically with energy.

Degeneracy in a continuum can be quantitatively represented by a *spectral density function* $g(E)$, such that there are $g(E)\,\mathrm{d}E$ distinct quantum states in the energy range E to $E + \mathrm{d}E$. To calculate $g(E)$, it is most convenient to make use of box normalization in three dimensions. Using the analog of (5.2.9) for each component of the wavevector, the energy (5.7.7) can be expressed

$$E_{\mathbf{n}} = \frac{h^2}{2mL^2}(n_1^2 + n_2^2 + n_3^2). \tag{5.8.1}$$

Each state is thereby specified by three integers $n_1 n_2 n_3$, which can be arrayed in a simple cubic lattice. In **n**-space there is, on the average, one state per unit volume. Now (5.8.1) is the equation of a sphere with radius

$$R = (2mE)^{\frac{1}{2}}L/h. \tag{5.8.2}$$

The total number of states with energies between 0 and E is hence equal to the volume of the sphere:

$$N_E = \frac{4\pi}{3}R^3 = \frac{4\pi L^3}{3h^3}(2mE)^{\frac{3}{2}}. \tag{5.8.3}$$

The derivative of this quantity w.r.t. E can thereby be identified with the density function:

$$g(E) = \frac{4\pi mV}{h^3}(2mE)^{\frac{1}{2}} = \frac{4\pi mV}{(2\pi)^3\hbar^2}k \tag{5.8.4}$$

having introduced $V = L^3$, the volume per particle. In a later application, we will require the spectral density *per unit solid angle*. Since (5.8.3) is independent of the momentum direction θ, ϕ, we simply divide by 4π to obtain

$$g(E, \theta, \phi) = \frac{mV}{(2\pi)^3\hbar^2}k. \tag{5.8.5}$$

The density of states can also be formulated in terms of wavenumber. Thus, using (5.7.3) in (5.8.3), we obtain

$$N_k = \frac{Vk^3}{6\pi^2} \tag{5.8.6}$$

representing the number of free-particle states with wavenumber between 0 and k. Such a description finds application in the theory of many-fermion systems, including metallic electrons and nuclear matter. Suppose an assembly of n spin $\frac{1}{2}$ fermions per unit volume fills a continuous set of one-particle quantum states in accordance with the exclusion principle. Taking account of the twofold spin degeneracy for each k, we have in the ground state of the system

$$2N_k = nV. \tag{5.8.7}$$

The boundary in wavevector space enclosing these occupied quantum states is called the *Fermi surface*. For free-particle eigenstates, as above, the Fermi surface is simply a sphere of radius

$$k_F = (3\pi^2 n)^{\frac{1}{3}}. \tag{5.8.8}$$

The corresponding values of p and E:

$$p_F = \hbar k_F, \qquad E_F = \hbar^2 k_F^2/2m \tag{5.8.9}$$

are called, respectively, the Fermi momentum and Fermi energy. Equation (5.8.8) applied to metals corresponds to the simple free-electron model. In real metals, the Fermi surface has a considerably more complicated geometrical structure.

It is also of interest to reexpress (5.8.3) in terms of momentum, using (5.1.10). The total number of states with momenta between 0 and p in magnitude is accordingly

$$N_p = \frac{4\pi V p^3}{3h^3}. \tag{5.8.10}$$

Note however that

$$\Delta\tau_{\mathbf{p}} = \frac{4\pi}{3} p^3 \tag{5.8.11}$$

represents the element of momentum space spanning these same values. Similarly

$$\Delta\tau_{\mathbf{r}} = V \tag{5.8.12}$$

represents the volume element in configuration space accessible to the particle. The number of quantum states contained within an element of *phase space* is therefore given by

$$\Delta N = h^{-3}\Delta\tau_{\mathbf{p}}\Delta\tau_{\mathbf{r}}. \tag{5.8.13}$$

This implies that one quantum state is associated, on the average, with each cell in phase space of extension

$$\Delta\tau_{\mathbf{p}}\Delta\tau_{\mathbf{r}} = h^3. \tag{5.8.14}$$

Whereas a classical system can be precisely located in phase space at every instant, its quantum analog is irreducibly smeared out. Given this interpretation, (5.8.14) becomes equivalent to the Heisenberg uncertainty principle (3.6.23).

Since the end result of computations must be independent of the method of counting quantum states, eqn (5.8.13) in differential form must be related to the energy degeneracy by

$$g(E)\,\mathrm{d}E = h^{-3}\,\mathrm{d}\tau_{\mathbf{p}}\,\mathrm{d}\tau_{\mathbf{r}}. \tag{5.8.15}$$

This equivalence is shown quite strikingly by independent calculation of the free-particle partition function in statistical mechanics. Using (5.8.4),

$$q \equiv \int_0^\infty g(E)\,\mathrm{e}^{-E/kT}\,\mathrm{d}E = (2\pi mkT)^{\frac{3}{2}}V/h^3. \tag{5.8.16}$$

On the other hand, by evaluation of the classical phase integral,

$$q = h^{-3}\iint \mathrm{e}^{-p^2/2mkT}\,\mathrm{d}\tau_{\mathbf{p}}\,\mathrm{d}\tau_{\mathbf{r}} = (2\pi mkT)^{\frac{3}{2}}V/h^3. \tag{5.8.17}$$

5.9 Spherical Waves

The free-particle Schrödinger equation (5.7.1) is also separable in spherical polar coordinates (in fact, in a total of 11 different orthogonal coordinate systems). By appropriate transformation of the Laplacian, we obtain

$$\left[\frac{1}{r^2}\frac{\partial}{\partial r}r^2\frac{\partial}{\partial r} + \frac{1}{r^2\sin\theta}\frac{\partial}{\partial\theta}\sin\theta\frac{\partial}{\partial\theta} + \frac{1}{r^2\sin^2\theta}\frac{\partial^2}{\partial\phi^2} + k^2\right]\psi(r,\theta,\phi) = 0. \tag{5.9.1}$$

Let us first consider spherically symmetrical solutions $\psi = \psi(r)$, representing states of zero angular momentum. The Schrödinger equation can then be rearranged to

$$\frac{\mathrm{d}^2}{\mathrm{d}r^2}(r\psi) + k^2 r\psi = 0. \tag{5.9.2}$$

This has solutions for $r\psi$ analogous to those of the one-dimensional free particle. The eigenfunctions for $k > 0$ are doubly degenerate and thus can be represented in alternative ways:

$$\psi_k(r) = \frac{\sin kr}{r} \quad \text{and} \quad \frac{\cos kr}{r} \tag{5.9.3}$$

or

$$\psi_k^{(\pm)}(r) = \frac{\mathrm{e}^{\pm ikr}}{r}. \tag{5.9.4}$$

These correspond, respectively, to stationary and to travelling waves. By appending the time factors to (5.9.4) we find

$$\Psi_k^{(+)}(r, t) = \frac{e^{i(kr-\omega t)}}{r} \tag{5.9.5}$$

and

$$\Psi_k^{(-)}(r, t) = \frac{e^{-i(kr+\omega t)}}{r} \tag{5.9.6}$$

which have the form of outgoing and incoming spherical waves, respectively.

The more general solutions to (5.9.1) are separable in the form

$$\psi(r, \theta, \phi) = R(r)Y(\theta, \phi) \tag{5.9.7}$$

where $Y(\theta, \phi)$ are spherical harmonics, which satisfy the angular momentum eigenvalue equation

$$-\left[\frac{1}{\sin\theta}\frac{\partial}{\partial\theta}\sin\theta\frac{\partial}{\partial\theta} + \frac{1}{\sin^2\theta}\frac{\partial}{\partial\phi^2}\right]Y_{lm}(\theta, \phi) = l(l+1)Y_{lm}(\theta, \phi). \tag{5.9.8}$$

The radial equation becomes

$$R''(r) + \frac{2}{r}R'(r) + \left[k^2 - \frac{l(l+1)}{r^2}\right]R(r) = 0. \tag{5.9.9}$$

Its solutions are spherical Bessel and Neumann functions†, $j_l(kr)$ and $n_l(kr)$, which have the asymptotic forms as $r \to \infty$:

$$j_l(kr) \sim \frac{\sin(kr - l\pi/2)}{kr} \tag{5.9.10}$$

and

$$n_l(kr) \sim -\frac{\cos(kr - l\pi/2)}{kr}. \tag{5.9.11}$$

Alternative solutions are the linear combinations

$$h_l^{(\pm)}(kr) = j_l(kr) \pm in_l(kr), \tag{5.9.12}$$

known as Hankel functions. In the asymptotic region,

$$h_l^{(\pm)}(kr) \sim \frac{e^{\pm ikr}}{kr} \tag{5.9.13}$$

For $l = 0$, the solutions to (5.9.9) reduce to those of (5.9.2).‡

† For details on the properties of spherical Bessel functions, see P. M. Morse and H. Feshbach, "Methods of Theoretical Physics", McGraw-Hill, New York, 1953, Chapter 5.

‡ Only the solutions with $j_l(kr)$ (including $j_0(kr) = \sin kr/kr$) are regular at $r = 0$. The others are, more precisely, solutions of

$$[\nabla^2 + k^2 + 4\pi\delta(\mathbf{r})]\,\psi(\mathbf{r}) = 0$$

(cf. Section 6.1).

We shall be most interested in those solutions which have the character of *outgoing* spherical waves, namely

$$\psi_{klm}^{(+)}(r, \theta, \phi) = h_l^{(+)}(kr) Y_{lm}(\theta, \phi). \tag{5.9.14}$$

Their asymptotic form can be represented by

$$\psi^{(+)}(r, \theta, \phi) = r^{-1} \mathrm{e}^{ikr} f(\theta, \phi) \quad (r \to \infty). \tag{5.9.15}$$

This asymptotic structure pertains, in fact, even in the presence of a short-range potential $V(\mathbf{r})$, one which falls off *more* rapidly than r^{-1}.

The probability density associated with the wavefunction (5.9.15) is

$$\rho(\mathbf{r}) = r^{-2} |f(\theta, \phi)|^2, \tag{5.9.16}$$

which falls off as the inverse square, in the characteristic fashion of spherical waves. The expression for current density is somewhat complicated in spherical coordinates. Since

$$\nabla = \mathbf{u}_r \frac{\partial}{\partial r} + \frac{\mathbf{u}_\theta}{r} \frac{\partial}{\partial \theta} + \frac{\mathbf{u}}{r \sin \theta} \frac{\partial}{\partial \phi} \tag{5.9.17}$$

we find

$$\mathbf{j} = \frac{\hbar k}{mr^2} |f(\theta, \phi)|^2 \mathbf{u}_r + \frac{\hbar^2}{2imr^3} \left(f^* \frac{\partial f}{\partial \theta} - f \frac{\partial f^*}{\partial \theta} \right) \mathbf{u}_\theta + \frac{\hbar^2}{2imr^3 \sin \theta} \left(f^* \frac{\partial f}{\partial \phi} - f \frac{\partial f^*}{\partial \phi} \right) \mathbf{u}_\phi. \tag{5.9.18}$$

The first term is the radial current while the remaining two represent circulations. The latter, in any event, fall off more rapidly by an inverse power of r. The dominant term in the asymptotic region is hence the radial component

$$j_r = \frac{\hbar k}{mr^2} |f(\theta, \phi)|^2. \tag{5.9.19}$$

Current density has dimensions of particle flux per unit area per unit time. For spherical waves it is convenient to define an alternative current density $j(\theta, \phi)$ representing the particle flux per unit *solid angle* per unit time. Since the subtended area per unit solid angle is simply r^2, we find

$$j(\theta, \phi) = \frac{\hbar k}{m} |f(\theta, \phi)|^2. \tag{5.9.20}$$

5.10 Scattering

In a prototype scattering experiment, particles in a narrow beam impinge upon a fixed target to be deflected at various angles. By appropriate detectors, the angular distribution (and perhaps the energy) of the scattered particles is measured. For purposes of theoretical analysis, we idealize the scattering process as a monoenergetic plane wave interacting with a fixed centre of force, giving rise to a secondary outgoing spherical wave. At macroscopic distance from the scattering centre, so that asymptotic forms apply, the wavefunction can be represented as a superposition of incident and scattered waves:

$$\psi(\mathbf{r}) = e^{i\mathbf{k}.\mathbf{r}} + r^{-1} e^{ikr} f(\theta, \phi), \tag{5.10.1}$$

known as the Faxén–Holtsmark formula.† The coordinate origin is, of course, chosen at the scattering centre. The incident plane wave has been simply normalized to contain one particle per unit volume. The function $f(\theta, \phi)$ is here called the *scattering amplitude.*

The Faxén-Holtsmark formula applies, more specifically to *elastic scattering*, in which the internal state of the target particle is not changed. Correspondingly the energy of the scattered particle is conserved. Thus the common factors $e^{-i\omega t}$ can be suppressed and the problem represented by time-independent equations.‡ Equation (5.10.1) is, in fact, an asymptotic solution of the time-independent Schrödinger equation

$$\left\{-\frac{\hbar^2}{2m}\nabla^2 + V(\mathbf{r})\right\}\psi(\mathbf{r}) = E\psi(\mathbf{r}) \tag{5.10.2}$$

with boundary conditions appropriate to the scattering geometry. Here the suitably short-range potential function $V(\mathbf{r})$ represents the interaction of the incident particles with the target.

By a simple modification, the above theory can be made applicable as well to *inelastic collisions.* Suppose the target particle undergoes a transition from initial state $\phi_0(q)$ to one of a set or final states $\phi_n(q)$. Then (5.10.1) can be generalized to

$$\psi(\mathbf{r}, q) = e^{i\mathbf{k}_0.\mathbf{r}}\phi_0(q) + \sum_n r^{-1} e^{ik_n r} f_n(\theta, \phi)\phi_n(q) \tag{5.10.3}$$

subject to overall energy conservation, whereby

$$E_0 + \frac{\hbar^2 k_0^2}{2m} = E_n + \frac{\hbar^2 k_n^2}{2m}. \tag{5.10.4}$$

In extremely fast high-energy collisions, however, nonconservation of energy may occur with bounds determined by the uncertainty principle.

In a real process, the target particle must recoil in order to conserve overall momentum. This can be taken into account most conveniently by

† H. Faxén and J. Holtsmark, *Z. Physik.* **45**, 307 (1927).

‡ An alternative treatment of scattering by time-dependent perturbation theory is given in Section 7.7.

transforming from the *laboratory* to the *centre-of-mass* coordinate system, as discussed by most standard texts on quantum mechanics. Kinetic energy then pertains to the *relative* motion while m represents the appropriate *reduced mass*.

Experimental data on the angular distribution of scattered particles are most conveniently represented in terms of the *differential scattering cross-section*:

$$\sigma(\theta, \phi) \equiv \frac{j(\theta, \phi)}{j_{\text{inc}}}. \tag{5.10.5}$$

Here $j(\theta, \phi)$ is the scattered current (per unit solid angle) while j_{inc} is the magnitude of the incident current (per unit area). The differential scattering cross-section has dimensions of area per unit solid angle. It can, in fact, be visualized as an equivalent area placed normal to the incident beam such as to intercept the same number of particles as are scattered per unit solid in a given direction.

A theoretical expression for the differential cross-section follows from the Faxén–Holtsmark formula. The scattered current corresponding to the outgoing spherical wave is given by (5.9.20). For the plane wave $e^{i\mathbf{k}\cdot\mathbf{r}}$, we have simply

$$j_{\text{inc}} = \frac{\hbar k}{m}. \tag{5.10.6}$$

Therefore

$$\sigma(\theta, \phi) = |f(\theta, \phi)|^2. \tag{5.10.7}$$

The scattering amplitude thus determines the differential cross-section.

Integration over all scattering angles gives the *total scattering cross-section*†

$$\sigma_{\text{tot}} = \iint \sigma(\theta, \phi) \sin\theta \, d\theta \, d\phi. \tag{5.10.8}$$

Total cross-sections in atomic and molecular collision processes are typically of the order of 10^{-16} cm^2, comparable to the geometrical cross-section of an atom 1 Å in diameter. Differential cross-sections are correspondingly of the order of 10^{-17} or 10^{-18} cm^2/steradian.

For nuclear processes, total cross-sections of the order of 10^{-26} cm^2 and

† The total cross-section is also related to the forward scattering amplitude $f(0)$ by the *optical theorem*:

$$\sigma_{\text{tot}} = \frac{4\pi}{k} \operatorname{Im} f(0)$$

derived in many textbooks.

differential cross-sections of the order of 10^{-27} or 10^{-28} cm^2/steradian are typical. The unit 10^{-24} cm^2 is called a *barn*. Differential cross-sections are commonly given in millibarns/steradian.

Much of what is known of the nature and interactions of atoms, nuclei and elementary particles is the result of detailed analysis of scattering cross-sections and their dependence on energy, polarization, etc.

6

Green's Functions

The English mathematician George Green (1793–1841) developed some very original techniques for the solution of problems in electromagnetic theory and fluid mechanics. He first introduced the notion of potential—of which Green's functions are a limiting case. Techniques based on Green's functions are now an indispensible part of applied mathematics and theoretical physics. In this Chapter, we shall discuss configuration-space Green's functions and their applications in quantum mechanics.† As a preliminary, however, we shall take up some classical aspects of Green's function formalism.

6.1 Green's Functions in Potential Theory

Poisson's equation in electrostatics

$$\nabla^2\Phi = -4\pi\rho \tag{6.1.1}$$

determines the Coulomb potential produced by a charge distribution. More generally, in the Lorentz-gauge formulation of electrodynamics, one encounters the inhomogeneous wave equation

$$\Box\Phi = 4\pi\rho. \tag{6.1.2}$$

These and other classical potential problems can be usefully formulated in terms of Green's functions.

A. *Poisson's Equation*

The Green's function $G(\mathbf{r}, \mathbf{r}')$ is defined as the potential at point $\mathbf{r}$ caused by a unit source at point $\mathbf{r}'$. Thus the Green's function for Poisson's equation satisfies

$$\nabla^2 G(\mathbf{r}, \mathbf{r}') = -4\pi\delta(\mathbf{r} - \mathbf{r}'). \tag{6.1.3}$$

† Beyond our scope are recent extensions of Green's-function formalism in quantum field theory and statistical mechanics. See, for example, T. D. Schultz *Quantum Field Theory and the Many Body Problem* (Gordon and Breach, New York and London, 1964) and references given therein. See also J. Linderberg and Y. Öhrm *Propagators in Quantum Chemistry* (Academic Press, London, 1973).

Since ∇^2 is a linear operator, one can add up the contributions from each source element to give the total potential:

$$\Phi(\mathbf{r}) = \int G(\mathbf{r}, \mathbf{r}')\, \rho(\mathbf{r}')\, \mathrm{d}^3\mathbf{r}'. \tag{6.1.4}$$

The solution of the potential problem—or, at least, an integral representation thereto—can thus be constructed if the appropriate Green's function is known.

On physical grounds, we know that electrostatic interactions are *central fields*. The corresponding Green's function should therefore depend only on the magnitude of the relative displacement:

$$G(\mathbf{r}, \mathbf{r}') = G(R) \tag{6.1.5}$$

in which

$$\mathbf{R} \equiv \mathbf{r} - \mathbf{r}', \qquad R = |\mathbf{R}| = |\mathbf{r} - \mathbf{r}'|. \tag{6.1.6}$$

To demonstrate this mathematically, write

$$\nabla^2 G(\mathbf{r}, \mathbf{r}'') = -4\pi\delta(\mathbf{r} - \mathbf{r}'') \tag{6.1.7}$$

and replace $\mathbf{r}$ by $\mathbf{r} - \mathbf{r}'$. Accordingly,

$$\nabla^2 G(\mathbf{r} - \mathbf{r}', \mathbf{r}'') = -4\pi\delta(\mathbf{r} - \mathbf{r}' - \mathbf{r}''). \tag{6.1.8}$$

Now if $\mathbf{r}''$ is reduced to 0, we obtain

$$\nabla^2 G(\mathbf{r} - \mathbf{r}', 0) = -4\pi\delta(\mathbf{r} - \mathbf{r}'), \tag{6.1.9}$$

which must represent the same equation as (6.1.3). This demonstrates that

$$G(\mathbf{r}, \mathbf{r}') = G(\mathbf{R}). \tag{6.1.10}$$

The Green's function equation can accordingly be written

$$\nabla_{\mathbf{R}}^2 G(\mathbf{R}) = -4\pi\delta(\mathbf{R}) \tag{6.1.11}$$

in which $\nabla_{\mathbf{R}}^2$ is the Laplacian in terms of the relative coordinates. Under an inversion of the coordinate system, $\mathbf{R} \to -\mathbf{R}$, eqn (6.1.11) transforms to

$$\nabla_{\mathbf{R}}^2 G(-\mathbf{R}) = -4\pi\delta(\mathbf{R}) \tag{6.1.12}$$

since both the Laplacian and the deltafunction are invariant. Therefore

$$G(\mathbf{R}) = G(-\mathbf{R}) \tag{6.1.13}$$

showing that the Green's function can depend only on the *magnitude* of $\mathbf{R}$, which proves (6.1.5). A corollary is the *reciprocity relation*

$$G(\mathbf{r}, \mathbf{r}') = G(\mathbf{r}', \mathbf{r}) \tag{6.1.14}$$

representing symmetry w.r.t. interchange of "field point" and "source point".

Let us now explicitly derive the Poisson Green's function. We can begin by taking the 3-dimensional Fourier transform of eqn (6.1.11). Noting that (cf. Table B)

$$FT\frac{\partial G}{\partial x} = ik_x G(\mathbf{k}), \quad \text{etc.} \tag{6.1.15}$$

we find

$$-k^2 G(\mathbf{k}) = -4\pi(2\pi)^{-3}. \tag{6.1.16}$$

Therefore (cf. B.21)

$$G(R) = \int G(\mathbf{k})\,\mathrm{e}^{i\mathbf{k}\cdot\mathbf{R}}\,\mathrm{d}^3\mathbf{k} = \frac{1}{2\pi^2}\int \frac{\mathrm{e}^{i\mathbf{k}\cdot\mathbf{R}}}{k^2}\,\mathrm{d}^3\mathbf{k}. \tag{6.1.17}$$

The integration is most easily carried out by transforming to spherical polar coordinates in **k**-space, with **R** defining the polar axis (cf. A.42):

$$G(R) = \frac{1}{2\pi^2}\int_0^\infty \int_0^{2\pi} \int_0^{\pi} \mathrm{e}^{ikR\cos\theta}\sin\theta\,\mathrm{d}k\,\mathrm{d}\theta\,\mathrm{d}\phi. \tag{6.1.18}$$

Integration over the angles gives

$$G(R) = \frac{2}{\pi R}\int_0^\infty \frac{\sin kR}{k}\,\mathrm{d}k\,. \tag{6.1.19}$$

The integral equals $\pi/2$, independent of R, so that

$$G(R) = 1/R \tag{6.1.20}$$

or, in terms of the original coordinates,

$$G(\mathbf{r},\mathbf{r}') = \frac{1}{|\mathbf{r}-\mathbf{r}'|}\,. \tag{6.1.21}$$

This Green's function represents the electrostatic potential of a unit point charge, in agreement, of course, with Coulomb's law.

Putting (6.1.21) into (6.1.4), we obtain the solution to Poisson's equation, representing the potential due to a continuous distribution of charge:

$$\Phi(\mathbf{r}) = \int \frac{\rho(\mathbf{r}')}{|\mathbf{r}-\mathbf{r}'|}\,\mathrm{d}^3\mathbf{r}'. \tag{6.1.22}$$

Putting (6.1.21) into (6.1.3) gives the important identity

$$\nabla^2\frac{1}{|\mathbf{r}-\mathbf{r}'|} = -4\pi\delta(\mathbf{r}-\mathbf{r}'). \tag{6.1.23}$$

This can also be demonstrated directly as follows. It is easily verified that

$$\nabla^2 \frac{1}{R} = 0 \qquad \text{for} \quad R \neq 0\,. \tag{6.1.24}$$

Moreover, by integrating over a sphere $R =$ constant and applying the divergence theorem, we find

$$\int \nabla^2 \frac{1}{R}\, \mathrm{d}^3\mathbf{R} = \int \boldsymbol{\nabla} \frac{1}{R} \cdot \mathrm{d}\mathbf{S} = 4\pi R^2 \left(-\frac{R}{R^3} \right) = -4\pi \tag{6.1.25}$$

showing that $\nabla^2\, 1/R$ behaves as $-4\pi\delta(\mathbf{R})$.

B. *General Solution*

To the solution (6.1.22) of Poisson's equation might be added some function $\Phi_0(\mathbf{r})$ which satisfies the corresponding homogeneous equation (Laplace's equation)

$$\nabla^2 \Phi_0(\mathbf{r}) = 0\,. \tag{6.1.26}$$

This function could be chosen so as to fulfil specified boundary conditions on a general solution to Poisson's equation:

$$\Phi(\mathbf{r}) = \int G(\mathbf{r}, \mathbf{r}')\, \rho(\mathbf{r}')\, \mathrm{d}^3\mathbf{r}' + \Phi_0(\mathbf{r}). \tag{6.1.27}$$

In the parlance of inhomogeneous differential equations, the two parts of (6.1.27) are, respectively, the *particular integral* and *complementary function.*

A more formal way to arrive at (6.1.27) makes use of *Green's theorem*: for appropriately well-behaved functions $u(\mathbf{r})$ and $v(\mathbf{r})$ within a region of volume V,

$$\int_V (u\nabla^2 v - v\nabla^2 u)\, \mathrm{d}\tau = \int_S (u\boldsymbol{\nabla} v - v\,\boldsymbol{\nabla} u) \cdot \mathrm{d}\mathbf{S} \tag{6.1.28}$$

in which S represents the closed surface enclosing V. Applying this to the functions $G(\mathbf{r}', \mathbf{r})$ and $\Phi(\mathbf{r}')$, we can write

$$\int_V [G(\mathbf{r}', \mathbf{r})\, \nabla'^2 \Phi(\mathbf{r}') - \Phi(\mathbf{r}')\, \nabla'^2 G(\mathbf{r}', \mathbf{r})]\, \mathrm{d}^3\mathbf{r}'$$
$$= \int_S [G(\mathbf{r}', \mathbf{r})\, \boldsymbol{\nabla}'\Phi(\mathbf{r}') - \Phi(\mathbf{r}')\, \boldsymbol{\nabla}' G(\mathbf{r}'\mathbf{r})] \cdot \mathrm{d}\mathbf{S}. \tag{6.1.29}$$

Making use of (6.1.1) and (6.1.3) in the first integral, we find, for points $\mathbf{r}$ within the volume V,

$$\Phi(\mathbf{r}) = \int_V G(\mathbf{r}, \mathbf{r}')\, \rho(\mathbf{r}')\, d^3\mathbf{r}'$$

$$+ \int_S [G(\mathbf{r}, \mathbf{r}_s)\, \nabla_s \Phi(\mathbf{r}_s) - \Phi(\mathbf{r}_s)\, \nabla_s G(\mathbf{r}, \mathbf{r}_s)] \cdot d\mathbf{S}. \qquad (6.1.30)$$

The surface integral is determined by the values of the potential and its normal gradient along the surface S. We have thus obtained an explicit representation for the complementary function $\Phi_0(\mathbf{r})$ in (6.1.27).

The general solution (6.1.30) reduces to (6.1.22) when the surface S is removed to infinity and the following boundary conditions imposed

$$\lim_{r\to\infty} \Phi(\mathbf{r}) = 0, \qquad \lim_{r\to\infty} \nabla\, \Phi(\mathbf{r}) = 0. \qquad (6.1.31)$$

Physically this means that explicit account is to be taken of all the charges in the system: should $\rho(\mathbf{r})$ vanish everywhere, then $\Phi(\mathbf{r})$ must likewise.

C. *Inhomogeneous Helmholtz Equation*

As a preliminary to the inhomogeneous wave equation (6.1.2), we shall work out the Green's function satisfying

$$(\nabla^2 + k^2)\, G(\mathbf{r}, \mathbf{r}', k) = -4\pi\delta(\mathbf{r} - \mathbf{r}'). \qquad (6.1.32)$$

Just as in the case of Poisson's equation, we have central field symmetry and

$$(\nabla_{\mathbf{R}}^2 + k^2)\, G(R, k) = -4\pi\delta(\mathbf{R}). \qquad (6.1.33)$$

The corresponding homogeneous equation (for $R \neq 0$) is identical to the free-particle Schrödinger equation (5.7.2). Correspondingly we find central-field solutions of the form $R^{-1}\, e^{\pm ikR}$. Indeed, by direct substitution in (6.1.32), using (6.1.23) to get the delta-function, it is verified that these solutions represent the required Green's function:

$$G^{(\pm)}(\mathbf{r}, \mathbf{r}', k) = \frac{e^{\pm ik|\mathbf{r}-\mathbf{r}'|}}{|\mathbf{r} - \mathbf{r}'|}. \qquad (6.1.34)$$

It is instructive nonetheless to proceed more methodically to this result. In analogy with our derivation of the Poisson Green's function, we take the Fourier transform of (6.1.33) (changing k to k_0 so that k can be used for the conjugate variable):

$$-(k^2 - k_0^2)\, G(\mathbf{k}) = -4\pi(2\pi)^{-3}. \qquad (6.1.35)$$

Thus

$$G(R, k) = \frac{1}{2\pi^2} \int \frac{e^{i\mathbf{k}\cdot\mathbf{R}}}{k^2 - k_0^2}\, d^3\mathbf{k}. \qquad (6.1.36)$$

Integrating over angles in **k**-space, as before, we obtain

$$G(R, k) = \frac{1}{i\pi R}\int_0^\infty \frac{k(e^{ikR} - e^{-ikR})}{k^2 - k_0^2}\,dk = \frac{1}{i\pi R}\int_{-\infty}^\infty \frac{k\,e^{ikR}}{k^2 - k_0^2}\,dk\,. \tag{6.1.37}$$

The integral, however, does not exist, owing to the singularities at $k = \pm k_0$. This difficulty can be circumvented by reinterpreting k_0 as a complex variable

$$k_0 \to k_0 + i\delta \qquad \delta > 0 \tag{6.1.38}$$

where δ is a small positive real number, eventually to be taken to the limit

$$\delta \to 0^+. \tag{6.1.39}$$

This strategem shifts the poles of the integrand off the real axis, to the points $k = \pm(k_0 + i\delta)$ indicated by ● in Figure 6.1. The resulting integral

$$I = \int_{-\infty}^\infty \frac{k\,e^{ikR}\,dk}{(k + k_0 + i\delta)(k - k_0 - i\delta)} \tag{6.1.40}$$

is most easily evaluated by closing a contour with a semicircle in the upper half plane, as shown in the figure. As the radius is taken to infinity, the integral

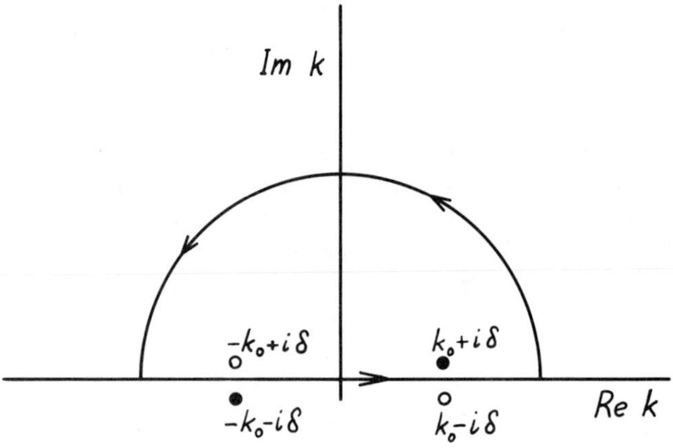

Fig. 6.1. Contour for evaluating Green's function for Helmholtz equation.

along the semicircle, where Im $k > 0$, goes to zero because of the exponential factor. Therefore, by the residue theorem,

$$I = \oint \frac{k\,e^{ikR}\,dk}{(k + k_0 + i\delta)(k - k_0 - i\delta)} = 2\pi i\mathscr{R}(k_0 + i\delta)\,. \tag{6.1.41}$$

In the limit (6.1.39), the residue reduces to

$$\mathscr{R}(k_0 + i0^+) = \frac{k_0\, e^{ik_0R}}{2k_0} \tag{6.1.42}$$

so that we obtain

$$G^{(+)}(R, k_0) = \frac{e^{ik_0R}}{R}. \tag{6.1.43}$$

Alternatively, by the substitution

$$k_0 \to k_0 - i\delta \qquad \delta > 0 \tag{6.1.44}$$

the poles are shifted to the opposite side of the real axis, as indicated by ○ in the figure. The same contour then gives

$$I = 2\pi i \mathscr{R}(-k_0 + i\delta) \tag{6.1.45}$$

resulting in

$$G^{(-)}(R, k_0) = \frac{e^{-ik_0R}}{R}. \tag{6.1.46}$$

It follows that G is a two-valued function of the complex variable k_0 with a branch cut along the real axis.

Whether the branch $G^{(+)}$ or $G^{(-)}$ is appropriate is determined by the physical situation to be represented.†

D. *Inhomogeneous Wave Equation*

This equation requires Green's functions which depend on time as well as space coordinates. The defining relation for the Green's function takes the form‡

$$\left(\nabla^2 - \frac{1}{c^2}\frac{\partial^2}{\partial t^2}\right) G(\mathbf{r}t, \mathbf{r}'t') = -4\pi\delta(\mathbf{r} - \mathbf{r}')\,\delta(t - t'). \tag{6.1.47}$$

Again we can demonstrate that G depends on the coordinates only through the variable R. Analogously, all time dependence enters through the variable

$$\tau \equiv t - t' \tag{6.1.48}$$

Accordingly, (6.1.47) simplifies to

$$\left(\nabla_{\mathbf{R}}^2 - \frac{1}{c^2}\frac{\partial^2}{\partial \tau^2}\right) G(R, \tau) = -4\pi\delta(\mathbf{R})\,\delta(\tau). \tag{6.1.49}$$

† Note that these two functions differ by a solution to the homogeneous Helmholtz equation:

$$G^{(+)} - G^{(-)} = 2i\frac{\sin k_0 R}{R}.$$

‡ In covariant notation

$$\Box G(x, x') = -4\pi i c\delta(x - x').$$

It is convenient to represent this Green's function as a Fourier integral in the time variable:

$$G(R, \tau) = \int_{-\infty}^{\infty} G(R, \omega)\, e^{-i\omega\tau}\, d\omega\,. \tag{6.1.50}$$

Fourier transformation of (6.1.49) accordingly results in

$$\left(\nabla^2 + \frac{\omega^2}{c^2}\right) G(R, \omega) = -4\pi(2\pi)^{-1}\delta(\mathbf{R}) \tag{6.1.51}$$

which has the form of the inhomogeneous Helmholtz equation (6.1.32). Corresponding to (6.1.34), we find the solutions

$$G^{(\pm)}(R, \omega) = \frac{e^{\pm i\omega R/c}}{2\pi R} \tag{6.1.52}$$

so that, by (6.1.50),

$$G^{(\pm)}(R, \tau) = \frac{1}{2\pi R}\int_{-\infty}^{\infty} e^{-i\omega(\tau \pm R/c)}\, d\omega\,. \tag{6.1.53}$$

The integral represents a deltafunction (cf. A.15), so finally

$$G^{(\pm)}(R, \tau) = \frac{\delta(\tau \mp R/c)}{R}\,. \tag{6.1.54}$$

The upper sign of this Green's function is in accord with classical notions of cause and effect: the potential at the field point $\mathbf{r}$ at time t is caused by the source point $\mathbf{r}'$ at an *earlier* time $t' = t - R/c$. The potential is thus *retarded* by the time interval it takes a light pulse of speed c to travel the intervening distance R. The *retarded* (or *causal*) *Green's function*

$$G^{(+)}(\mathbf{r}t, \mathbf{r}'t') = \frac{\delta[t - t' - (R/c)]}{R} \tag{6.1.55}$$

thus represents the solution to (6.1.47) in classical electrodynamics.† The solution to the inhomogeneous wave equation (6.1.2) is accordingly the *retarded potential*

$$\Phi(\mathbf{r}, t) = \iint \frac{\rho(\mathbf{r}'t')}{R}\,\delta\left(t - t' - \frac{R}{c}\right) d^3\mathbf{r}'\, dt' \tag{6.1.56}$$

By integration over t', this can be expressed

$$\Phi(\mathbf{r}, t) = \int \frac{[\rho(\mathbf{r}'t')]}{R}\, d^3\mathbf{r}' \tag{6.1.57}$$

† *Advanced potentials*, in which $t' = t + (R/c)$, are also mathematically valid solutions of the wave equations. They are invoked in completely covariant classical electrodynamics and in connection with time-reversal symmetry in quantum mechanics (cf. Section 4.7).

in terms of the *retarded charge density*

$$[\rho(\mathbf{r}', t')] \equiv \rho\left(\mathbf{r}', t - \frac{R}{c}\right). \tag{6.1.58}$$

Analogous retarded solutions apply to the vector potential and the Hertzian vector, viz,

$$\mathbf{A}(\mathbf{r}, t) = \frac{1}{c}\int \frac{[\mathbf{j}(\mathbf{r}', t')]}{R}\, \mathrm{d}^3\mathbf{r}' \tag{6.1.59}$$

and

$$\mathbf{\Pi}(\mathbf{r}, t) = \int \frac{[\mathbf{P}(\mathbf{r}', t')]}{R}\, \mathrm{d}^3\mathbf{r}'. \tag{6.1.60}$$

6.2 General Formalism of Green's Functions and Operators

Consider as the prototype problem an inhomogeneous operator equation

$$\mathscr{A}\psi(q) = C\chi(q) \tag{6.2.1}$$

to be solved for the unknown function $\psi(q)$. Here $\chi(q)$ is a known function and C some convenient constant, generally compounded of factors such as -1, 4π, i, $\hbar$, c, etc. The theory is easily generalized, if necessary, to time-dependent operators and functions. The Green's function for the operator $\mathscr{A}$ is defined as the solution of the corresponding equation for a "unit source" inhomogeneity:

$$\mathscr{A}G(q, q') = C\delta(q - q'). \tag{6.2.2}$$

It follows that the general solution to the inhomogeneous equation (6.2.1) takes the form

$$\psi(q) = \int G(q, q')\,\chi(q')\,\mathrm{d}\tau' + \psi_0(q) \tag{6.2.3}$$

in which $\psi_0(q)$ satisfies the homogeneous equation

$$\mathscr{A}\psi_0(q) = 0 \tag{6.2.4}$$

in accordance with the required limiting conditions.

Equation (6.2.2) can be reformulated as an operator relation

$$\mathscr{A}\mathscr{G} = C\mathbb{1} \tag{6.2.5}$$

in which $\mathbb{1}$ is the identity operator and $\mathscr{G}$, the *Green's operator* or *resolvent*. Since

$$\mathscr{G} = C\mathscr{A}^{-1} \tag{6.2.6}$$

$\mathscr{G}$ is essentially the inverse operator to $\mathscr{A}$ (assuming, for the moment, that $\mathscr{A}^{-1}$ does exist). The connection between the Green's operator and function is shown by applying (6.2.5) to a deltafunction:

$$\mathscr{A}\mathscr{G}\delta(q - q') = C\delta(q - q'). \tag{6.2.7}$$

Comparing with (6.2.2) shows the formal relationship

$$\mathscr{G}\delta(q - q') = G(q, q'). \tag{6.2.8}$$

More concretely, by integrating the product of (6.2.8) with an arbitrary function $\chi(q')$, we obtain

$$\mathscr{G}\chi(q) = \int G(q, q')\,\chi(q')\,\mathrm{d}\tau'. \tag{6.2.9}$$

This identifies the Green's function as the integral kernel representing the Green's operator (cf. 3.3.7).

A. *Self-adjoint and Unitary Operators*

Of particular importance in applications are the special properties of Green's functions for these types of operators. First consider the case in which $\mathscr{A}$ is self-adjoint (hermitian). Applying the integral relation which defines hermiticity to $G(q, q')$ and $G(q, q'')$, we find

$$\int G^*(q, q')\,\mathscr{A}G(q, q'')\,\mathrm{d}\tau = \int [\mathscr{A}G(q, q')]^*\,G(q, q'')\,\mathrm{d}\tau. \tag{6.2.10}$$

Using (6.2.2) and integrating over the deltafunctions, we obtain

$$G^*(q'', q') = G(q', q'') \tag{6.2.11}$$

a generalization of the reciprocity relation (6.1.14). This also follows directly from (6.2.6) when $\mathscr{A}$, hence $\mathscr{G}$, is self-adjoint. The general solution (6.2.3) can also be formally demonstrated from the hermitian property. Consider the identity

$$\int G^*(q', q)\,\mathscr{A}'[\psi(q') - \psi_0(q')]\,\mathrm{d}\tau' = \int [\mathscr{A}'G(q', q)]^*\,[\psi(q') - \psi_0(q')]\,\mathrm{d}\tau'. \tag{6.2.12}$$

Applying (6.2.1) and (6.2.4) to the left-hand side and (6.2.2) on the right, we obtain

$$\int G^*(q', q)\,\chi(q')\,\mathrm{d}\tau' = \int \delta(q - q')\,[\psi(q') - \psi_0(q')]\,\mathrm{d}\tau' \tag{6.2.13}$$

which, by virtue of (6.2.11), reduces to (6.2.3).

For a unitary operator, which we shall designate $\mathscr{U}$, it is most convenient to set $C = 1$. The inhomogeneous equation thereby reads

$$\mathscr{U}\psi(q) = \chi(q). \tag{6.2.14}$$

The Green's function $K(q, q')$ is accordingly determined by

$$\mathscr{U}K(q, q') = \delta(q - q'). \tag{6.2.15}$$

We first show that there exists no non-trivial solution of the inhomogeneous equation

$$\mathscr{U}\psi_0(q) = 0. \tag{6.2.16}$$

Obviously,

$$\int [\mathscr{U}\psi_0(q)]^* \mathscr{U}\Psi_0(q)\, \mathrm{d}\tau = 0. \tag{6.2.17}$$

But by the definition of the adjoint and the unitary property, the last integral transforms to

$$\int \psi_0^*(q)\, \mathscr{U}^\dagger \mathscr{U}\psi_0(q)\, \mathrm{d}\tau = \int |\psi_0(q)|^2\, \mathrm{d}\tau = 0 \tag{6.2.18}$$

which implies that

$$\psi_0(q) = 0. \tag{6.2.19}$$

Corresponding to (6.2.15) is the Green's operator equation

$$\mathscr{U}\mathscr{K} = \mathbb{1} \tag{6.2.20}$$

which immediately identifies

$$\mathscr{K} = \mathscr{U}^\dagger. \tag{6.2.21}$$

One can also define a Green's function for the adjoint operator $\mathscr{U}^\dagger$:

$$\mathscr{U}^\dagger K^\dagger(q, q') = \delta(q - q'). \tag{6.2.22}$$

Clearly, the Green's operator for $\mathscr{U}^\dagger$ is $\mathscr{U}$. It is not difficult to show that

$$\int K^*(q, q')\, K(q, q'')\, \mathrm{d}\tau = \delta(q - q') \tag{6.2.23}$$

and that

$$K^*(q, q') = K^\dagger(q', q). \tag{6.2.24}$$

Finally, the solution to (6.2.14) follows from the identity

$$\int K(q, q')\, \mathscr{U}'\psi(q')\, \mathrm{d}\tau' = \int [\mathscr{U}'^\dagger K^*(q, q')]^* \psi(q')\, \mathrm{d}\tau'. \tag{6.2.25}$$

Using (6.2.14) on the left-hand side, (6.2.24) and (6.2.22) on the right, we conclude that

$$\psi(q) = \int K(q, q')\, \chi(q')\, \mathrm{d}\tau' . \qquad (6.2.26)$$

B. *Spectral Representation*

To the operator $\mathscr{A}$ corresponds a spectrum of eigensolutions satisfying

$$\mathscr{A}\phi_n(q) = a_n \phi_n(q). \qquad (6.2.27)$$

By virtue of (6.2.6), we find the associated eigenvalue equation for the Green's operator:

$$\mathscr{G}\phi_n(q) = C a_n^{-1} \phi_n(q). \qquad (6.2.28)$$

By the closure relation for the complete set of eigenfunctions

$$\underset{n}{\mathbf{S}}\, \phi_n^*(q')\, \phi_n(q) = \delta(q - q'). \qquad (6.2.29)$$

Putting this representation for the deltafunction into (6.2.8) and using (6.2.28), we obtain the *spectral representation* for the Green's function:

$$G(q, q') = C \underset{n}{\mathbf{S}}\, a_n^{-1} \phi_n^*(q')\, \phi_n(q). \qquad (6.2.30)$$

For hermitian operators, the eigenvalues a_n are real. For unitary operators, the eigenvalues have the structure

$$u_n = \mathrm{e}^{i\alpha_n} \qquad (6.2.31)$$

so that the spectral representation takes the form

$$K(q, q') = \underset{n}{\mathbf{S}}\, \mathrm{e}^{-i\alpha_n} \phi_n^*(q')\, \phi_n(q). \qquad (6.2.32)$$

The special properties of hermitian and unitary Green's functions can alternatively be deduced from these expansions.

The Green's function for an operator can be constructed if its eigenvalues and eigenfunctions are known and the sum (6.2.30) can be evaluated. As an illustration, we return to Poisson's equation, which corresponds to $\mathscr{A} = \nabla^2$, $C = -4\pi$. The Laplacian operator satisfies a continuum eigenvalue equation with

$$\phi_n(\mathbf{r}) = (2\pi)^{-\frac{3}{2}}\, \mathrm{e}^{i\mathbf{k}\cdot\mathbf{r}}, \qquad a_n = -k^2. \qquad (6.2.33)$$

Thus

$$G(\mathbf{r}, \mathbf{r}') = -4\pi(2\pi)^{-3} \int \frac{\mathrm{e}^{i\mathbf{k}\cdot(\mathbf{r}-\mathbf{r}')}}{-k^2}\, \mathrm{d}^3\mathbf{k} \qquad (6.2.34)$$

which is identical to (6.1.17).

The theory of this section is contingent upon the existence of the inverse operator $\mathscr{A}^{-1}$. Its failure to exist might arise, for example, if its spectrum includes the eigenvalue 0. In such cases, some related operator $\mathscr{A}(\alpha)$, in which α is a complex parameter, might still have an inverse. This was, in essence, done in the preceding section, the Helmholtz operator $\nabla^2 + k_0^2$ being replaced by $\nabla^2 + (k_0 + i\delta)^2$ to remove its singularities.

6.3 Green's Functions for the Schrödinger Equation

Consider a system described by the time-dependent Schrödinger equation

$$\left(i\hbar\frac{\partial}{\partial t} - \mathscr{H}_0\right)\Psi_0(q, t) = 0 \tag{6.3.1}$$

which we presume to have solved exactly. Suppose now that the system is subjected to a perturbation represented by the operator $\mathscr{V}(q, t)$. The perturbed Schrödinger equation

$$\left[i\hbar\frac{\partial}{\partial t} - \mathscr{H}_0 - \mathscr{V}(q, t)\right]\Psi(q, t) = 0 \tag{6.3.2}$$

can be formulated as an inhomogeneous equation by writing

$$\left(i\hbar\frac{\partial}{\partial t} - \mathscr{H}_0\right)\Psi(q, t) = \mathscr{V}(q, t)\,\Psi(q, t). \tag{6.3.3}$$

It can accordingly be solved in terms of a Green's function determined by

$$\left(i\hbar\frac{\partial}{\partial t} - \mathscr{H}_0\right)G_0(qt, q't') = \delta(q - q')\,\delta(t - t'). \tag{6.3.4}$$

The solution to (6.3.3) takes the form†

$$\Psi(q, t) = \Psi_0(q, t) + \int_q\int_{t=-\infty}^{\infty} G_0(qt, q't')\,\mathscr{V}(q', t')\,\Psi(q't')\,\mathrm{d}q'\,\mathrm{d}t'. \tag{6.3.5}$$

The Schrödinger differential equation has thereby been transformed into an integral equation with the Green's function as kernel. This last equation is a bit more complicated than the situation in the preceding section because the unknown function, being part of the inhomogeneity, occurs also in the integral.

Causal relationship requires that the state of the system at time t depend only on its history, but not on its future. Thus, if eqn (6.3.5) is to make sense physically, the integrand must vanish for $t' > t$. This is accomplished by means of a *causal* (or *retarded*) Green's function for which

$$G_0^{(+)}(qt, q't') = 0, \qquad t' > t. \tag{6.3.6}$$

† In the remainder of this Chapter we shall write $\mathrm{d}q'$ for the configuration element to avoid confusion with the time variable τ.

(In the relativistic theory, the Green's function must vanish for $t' > t + (R/c.)$ We shall see presently how this is constructed.

The unperturbed Hamiltonian $\mathscr{H}_0$ will always be taken as a time-independent operator. Consequently, the Green's function will contain time dependence only through

$$\tau = t - t' \tag{6.3.7}$$

and (6.3.4) can be recast as

$$\left(i\hbar\frac{\partial}{\partial\tau} - \mathscr{H}_0\right) G_0(q, q', \tau) = \delta(q - q')\,\delta(\tau). \tag{6.3.8}$$

Let us now Fourier transform this equation in the time variable, denoting the conjugate variable as $E/\hbar$. The result is

$$(E - \mathscr{H}_0)\, G_0(q, q', E) = \delta(q - q') \tag{6.3.9}$$

such that

$$G_0(q, q', \tau) = \frac{1}{2\pi}\int_{-\infty}^{\infty} G_0(q, q', E)\, \mathrm{e}^{-iE\tau/\hbar}\, \mathrm{d}E/\hbar\,. \tag{6.3.10}$$

Now $G_0(q, q', E)$ is the Green's function for the time-independent Schrödinger equation. Its spectral representation in terms of the eigenstates of $\mathscr{H}_0$ is given by (cf. 6.2.30)

$$G_0(q, q', E) = \mathop{\mathsf{S}}_n \frac{\psi_n^*(q')\,\psi_n(q)}{E - E_n}\,. \tag{6.3.11}$$

The energy Green's function is evidently singular at every point $E = E_n$ belonging to the eigenvalue spectrum of $\mathscr{H}_0$. Thus (6.3.10) is an improper integral unless the parameter E is extended into the complex plane. G_0 is however an analytic function of E at all other points in the complex plane. By the substitution

$$E \to E + i\varepsilon, \tag{6.3.12}$$

where ε is a small positive quantity to be taken in the limit

$$\varepsilon \to 0^+, \tag{6.3.13}$$

the poles of $G_0(q, q', E + i0^+)$ are shifted infinitesimally *below* the real axis. We proceed now to evaluate (6.3.10) by the contour technique used in Section 6.1C. For $\tau > 0$ we augment the real axis by an infinite semicircle in the *negative* imaginary half plane, such that the exponential tends to zero. By the residue theorem, the result is

$$G_0^{(+)}(q, q', \tau) = -\frac{i}{\hbar}\mathop{\mathsf{S}}_n \psi_n^*(q')\,\psi_n(q)\,\mathrm{e}^{-iE_n\tau/\hbar} \qquad (\tau > 0). \tag{6.3.14}$$

For $\tau < 0$, the contour must be closed with a semicircle in the *upper* half plane. Since no poles are thereby enclosed, the residue is zero and

$$G_0^{(+)}(q, q', \tau) = 0 \qquad \tau < 0. \tag{6.3.15}$$

Noting that $\tau < 0$ corresponds to $t' > t$, we conclude that $G_0^{(+)}(q, q', \tau)$ fulfils the causality condition (6.3.6). Combining (6.3.14) and (6.3.15) we can write

$$G_0^{(+)}(q, q', \tau) = -\frac{i}{\hbar}\theta(\tau)\, K(q, q', \tau) \tag{6.3.16}$$

where $\theta(\tau)$ is the unit step function (cf. A.18) and

$$K(q, q', \tau) \equiv \mathop{\mathbf{S}}_n \psi_n^*(q')\,\psi_n(q)\,\mathrm{e}^{-iE_n\tau/\hbar}. \tag{6.3.17}$$

Correspondingly, the causally correct energy Green's function is identified as

$$G_0^{(+)}(q, q', E) = G_0(q, q', E + i0^+). \tag{6.3.18}$$

This could all have been done more quickly in terms of Green's operators. The operator equation equivalent to (6.3.1) is

$$\left(i\hbar\frac{\partial}{\partial\tau} - \mathscr{H}_0\right)\mathscr{U}_0(\tau) = 0 \tag{6.3.19}$$

where $\mathscr{U}_0(\tau)$ is the evolution operator (cf. 4.8.16)

$$\mathscr{U}_0(\tau) = \mathrm{e}^{-i\mathscr{H}_0\tau/\hbar}. \tag{6.3.20}$$

Corresponding to (6.3.1) we have

$$\left(i\hbar\frac{\partial}{\partial\tau} - \mathscr{H}_0\right)\mathscr{G}_0(\tau) = \delta(\tau). \tag{6.3.21}$$

Comparing (6.3.21) with (6.3.19) and recalling that

$$\theta'(\tau) = \delta(\tau) \tag{6.3.22}$$

we deduce the relation

$$\mathscr{G}_0^{(+)}(\tau) = -\frac{i}{\hbar}\theta(\tau)\,\mathscr{U}_0(\tau). \tag{6.3.23}$$

The superscript $(+)$ is applied since this operator has the requisite causal behaviour, vanishing for $\tau < 0$. Since

$$-\theta'(-\tau) = \delta(\tau) \tag{6.3.24}$$

as well, an alternative solution to (6.3.21) gives the *advanced Green's operator*

$$\mathscr{G}_0^{(-)}(\tau) = \frac{i}{\hbar}\theta(-\tau)\,\mathscr{U}_0(\tau) \tag{6.3.25}$$

which represents backward evolution in time. By Fourier transformation of (6.3.21) we obtain

$$(E - \mathscr{H}_0)\mathscr{G}_0(E) = \mathbb{1} \tag{6.3.26}$$

where $\mathscr{G}_0(E)$ is the Green's operator for the time-independent Schrödinger equation:

$$\mathscr{G}_0(E) = (E - \mathscr{H}_0)^{-1}. \tag{6.3.27}$$

Also

$$\mathscr{G}_0(E) = \int_{-\infty}^{\infty} \mathscr{G}_0(\tau)\, e^{iE\tau/\hbar}\, d\tau. \tag{6.3.28}$$

To determine the causal Green's operator $\mathscr{G}_0^{(+)}(E)$ put (6.3.23) into (6.3.28):

$$\mathscr{G}_0^{(+)}(E) = -\frac{i}{\hbar}\int_{-\infty}^{\infty} \theta(\tau)\,\mathscr{U}_0(\tau)\, e^{iE\tau/\hbar}\, d\tau = -\frac{i}{\hbar}\int_0^{\infty} e^{i(E-\mathscr{H}_0)\tau/\hbar}\, d\tau. \tag{6.3.29}$$

Now, for the usual type of Hamiltonian, convergence of this integral can be guaranteed only if a factor $e^{-\varepsilon\tau/\hbar}$ is introduced by giving E a positive imaginary component as in (6.3.14). Integration then results in

$$\mathscr{G}_0^{(+)}(E) = (E + i0^+ - \mathscr{H}_0)^{-1} \tag{6.3.30}$$

in agreement with (6.3.18).

The retarded and advanced Green's operators obey the following conjugation relations:

$$\mathscr{G}_0^{(+)}(E)^\dagger = \mathscr{G}_0^{(-)}(E) \tag{6.3.31}$$

and

$$\mathscr{G}_0^{(+)}(\tau)^\dagger = \mathscr{G}_0^{(-)}(\tau). \tag{6.3.32}$$

For the corresponding Green's functions

$$G_0^{(+)}(q, q', E)^* = G_0^{(-)}(q', q, E) \tag{6.3.33}$$

and

$$G_0^{(+)}(q, q', \tau)^* = G_0^{(-)}(q', q, -\tau) \tag{6.3.34}$$

or, in terms of the original time variables,

$$G_0^{(+)}(qt, q't')^* = G_0^{(-)}(q't', qt). \tag{6.3.35}$$

We have limited our considerations to Green's functions in the coordinate representation. The reader should be aware, however, that a great deal of work is done using Green's functions and operators in *momentum space*, e.g., $G(pt, p't)$.

In what follows, we specialize to perturbations independent of time. We seek the solution of the perturbed Schrödinger equation

$$[\mathscr{H}_0 + \mathscr{V}(q)]\,\psi(q) = E\psi(q) \tag{6.3.36}$$

which reduces in the limit $\mathscr{V} \to 0$ to the unperturbed solution

$$\mathscr{H}_0\psi_0(q) = E\psi_0(q) \tag{6.3.37}$$

with the *same eigenvalue.* This corresponds to a time-independent description of an *elastic process.* But the theory is more generally applicable if the system is made sufficiently inclusive that its total energy is conserved. The preceding formalism is applied by arranging (6.3.36) in the form

$$(E - \mathscr{H}_0)\,\psi(q) = \mathscr{V}(q)\,\psi(q) \tag{6.3.38}$$

and making use of the causal energy Green's function satisfying

$$(E - \mathscr{H}_0)\,G_0^{(+)}(q, q', E) = \delta(q - q'). \tag{6.3.39}$$

The general solution is the *Lippmann-Schwinger integral equation*:

$$\psi(q) = \psi_0(q) + \int G_0^{(+)}(q, q', E)\,\mathscr{V}(q')\,\psi(q')\,\mathrm{d}q'. \tag{6.3.40}$$

In algebraic form

$$\psi(q) = \psi_0(q) + \mathscr{G}_0^{(+)}(E)\,\mathscr{V}(q)\,\psi(q). \tag{6.3.41}$$

The iterative solution, obtained by replacing $\psi(q')$ in the integrand by successive approximations containing only $\psi_0(q')$, results in the *Born series*:

$$\psi(q) = \{1 + \mathscr{G}_0^{(+)}(E)\,\mathscr{V}(q) + \mathscr{G}_0^{(+)}(E)\,\mathscr{V}(q)\,\mathscr{G}_0^{(+)}(E)\,\mathscr{V}(q) + \ldots\}\,\psi_0(q)$$

$$= \sum_{n=0}^{\infty} [\mathscr{G}_0^{(+)}(E)\,\mathscr{V}(q)]^n\psi_0(q). \tag{6.3.42}$$

Keeping only the first-order term, such that

$$\psi(q) \approx \psi_0(q) + \mathscr{G}_0^{(+)}(E)\,\mathscr{V}(q)\,\psi_0(q) \tag{6.3.43}$$

is called the *first Born approximation* (or simply *the* Born approximation). Higher order Born approximations can be obtained analogously.

It is possible, in principle, to sum the Born series explicitly. Noting that

$$E - \mathscr{H}_0 - \mathscr{V} = \mathscr{G}_0(E)^{-1} + \mathscr{V} \tag{6.3.44}$$

it is suggestive to define the *perturbed Green's operator*

$$\mathscr{G}(E) \equiv (E - \mathscr{H}_0 - \mathscr{V})^{-1} = (E - \mathscr{H})^{-1} \tag{6.3.45}$$

with the obvious modification for $\mathscr{G}^{(+)}(E)$. Now (6.3.44) can be written

$$\mathscr{G}_0(E)^{-1} = \mathscr{G}(E)^{-1} + \mathscr{V}. \tag{6.3.46}$$

Premultiplying by $\mathscr{G}_0(E)$ then postmultiplying by $\mathscr{G}(E)$ leads to the *Dyson equation*

$$\mathscr{G}(E) = \mathscr{G}_0(E) + \mathscr{G}_0(E)\mathscr{V}\mathscr{G}(E). \tag{6.3.47}$$

By iteration, we obtain

$$\mathscr{G} = \mathscr{G}_0 + \mathscr{G}_0\mathscr{V}\mathscr{G}_0 + \mathscr{G}_0\mathscr{V}\mathscr{G}_0\mathscr{V}\mathscr{G}_0 + \ldots = \sum_{n=0}^{\infty} (\mathscr{G}_0\mathscr{V})^n\mathscr{G}_0. \tag{6.3.48}$$

Thus, (6.3.42) can be summed explicitly to give

$$\psi(q) = \psi_0(q) + \mathscr{G}^{(+)}(E)\mathscr{V}(q)\,\psi_0(q), \tag{6.3.49}$$

a modification of the Lippmann–Schwinger equation in which the perturbed wavefunction does *not* appear in the integrand. In practice, however, this is no advantage unless the perturbed Green's function can be evaluated in closed form. Otherwise the infinite series merely reappears in the expression for $\mathscr{G}(E)$. The corresponding time-dependent Green's operators satisfy the relation

$$\mathscr{G}(t, t') = \mathscr{G}_0(t - t') + \int \mathscr{G}_0(t - t'')\mathscr{V}(t'')\mathscr{G}_0(t'' - t')\,\mathrm{d}t'' + \ldots \quad (t < t'' < t'). \tag{6.3.50}$$

For the associated Green's functions, written for brevity in terms of space-time variables $x = (q, t)$,

$$G(x, x') = G_0(x, x') + \int G_0(x, x'')\mathscr{V}(x'')\,G_0(x'', x')\,\mathrm{d}^4x'' + \ldots \tag{6.3.51}$$

We conclude this section with a brief comment on the physical significance of quantum-mechanical Green's functions. In potential theory, the meaning of, say $G(x, x')$ is fairly obvious: it represents the influence of a source point x' in space-time on a field point x. Quantum dynamics is concerned with the evolution of a system between times t' and t. Thus the Green's function can be interpreted as a *transition amplitude* connecting a pair of space-time points.

The "system" of most frequent concern to physicists is a free particle scattered by a potential. The Green's function $G(x, x')$ is accordingly called a *propagator*.† In terms of this notion, the series (6.3.51) can be verbalized as follows. The "true" propagator $G(x, x')$ is approximated to zeroth order by the *free propagator* $G_0(x, x')$. The first-order correction entails free propagation from x' to x'', "scattering" by the potential $\mathscr{V}$ at x'', followed by free propagation from x'' to x. Higher-order terms involve mutiple scattering.

† Various authors apply the term to *any* Green's function or operator.

Propagations and interactions can be represented on space-time plots called Feynman diagrams, which have provided a valuable technique for higher-order perturbation calculations in quantum field theory and statistical mechanics.

6.4 Application to Scattering Theory

Green's function formalism is most typically applied to problems of scattering. The simplest situation entails elastic scattering of "structureless" particles, in which case $\mathscr{H}_0$ represents the kinetic energy of an incident particle and $V(\mathbf{r})$, a short-range interaction potential. We must first obtain the causal free-particle Green's function $G_0^{(+)}(\mathbf{r}, \mathbf{r}', E)$. Specifically,

$$\mathscr{H}_0 = -\frac{\hbar^2}{2m}\nabla^2 \tag{6.4.1}$$

(where m is to be interpreted as the reduced mass of the interacting particles) so that the defining relation for the Green's function is

$$\left(E + i\varepsilon + \frac{\hbar^2}{2m}\nabla^2\right) G_0^{(+)}(\mathbf{r}, \mathbf{r}', E) = \delta(\mathbf{r} - \mathbf{r}'). \tag{6.4.2}$$

Since the free-particle energy has the form

$$E = \hbar^2 k^2/2m \tag{6.4.3}$$

(6.4.2) may be rearranged to

$$\frac{\hbar^2}{2m}[\nabla^2 + (k + i\delta)^2]\, G_0^{(+)}(\mathbf{r}, \mathbf{r}', E) = \delta(\mathbf{r} - \mathbf{r}') \tag{6.4.4}$$

in which

$$\varepsilon \approx \frac{k\hbar^2}{m}\delta\,. \tag{6.4.5}$$

Now the solution to (6.4.4) is, apart from a constant factor, just the Helmholtz Green's function for outgoing waves derived in Section 6.1C. Thus

$$G_0^{(+)}(\mathbf{r}, \mathbf{r}', E) = -\frac{m}{2\pi\hbar^2}\frac{e^{ik|\mathbf{r}-\mathbf{r}'|}}{|\mathbf{r} - \mathbf{r}'|}\,. \tag{6.4.6}$$

We can now set up the Lippmann–Schwinger equation:

$$\psi(\mathbf{r}) = e^{i\mathbf{k}\cdot\mathbf{r}} - \frac{m}{2\pi\hbar^2}\int \frac{e^{ik|\mathbf{r}-\mathbf{r}'|}}{|\mathbf{r} - \mathbf{r}'|} V(\mathbf{r}')\,\psi(\mathbf{r}')\,d^3\mathbf{r}' \tag{6.4.7}$$

in which the unperturbed state is represented as a plane wave of unit probability density. The scattering cross-section (cf. Section 5.10) is determined by the asymptotic behaviour of this solution as $r \to \infty$. The law of cosines implies the identity

$$|\mathbf{r} - \mathbf{r}'| = r\left(1 - \frac{2\mathbf{r}\cdot\mathbf{r}'}{r^2} + \frac{r'^2}{r^2}\right)^{\frac{1}{2}}. \tag{6.4.8}$$

We need consider values of r' only up to the effective range of the potential $\mathscr{V}(\mathbf{r}')$. Presumably, $r \gg r'$, so that

$$|\mathbf{r} - \mathbf{r}'| \approx r - \frac{\mathbf{r}\cdot\mathbf{r}'}{r} \tag{6.4.9}$$

neglecting terms of second order in r'/r. It is adequate to approximate

$$\frac{1}{|\mathbf{r} - \mathbf{r}'|} \approx \frac{1}{r}. \tag{6.4.10}$$

In the exponential, however, the first-order term of (6.4.9) will contribute a factor of order $e^{ikr'}$, which can not be discarded.† We write therefore

$$e^{ik|\mathbf{r}-\mathbf{r}'|} \approx e^{ikr}\, e^{-ik\mathbf{r}\cdot\mathbf{r}'/r}. \tag{6.4.11}$$

It is convenient to introduce $\mathbf{k}'$, the wavevector of a scattered particle. It has the magnitude k and the direction of $\mathbf{r}$, such that

$$\mathbf{k}' = k\mathbf{r}/r \tag{6.4.12}$$

in terms of which

$$e^{ik|\mathbf{r}-\mathbf{r}'|} \approx e^{ikr}\, e^{-i\mathbf{k}'\cdot\mathbf{r}'}. \tag{6.4.13}$$

By virtue of (6.4.10) and (6.4.13), (6.4.6) assumes the structure of the Faxén–Holtsmark formula (5.10.1):

$$\psi(\mathbf{r}) = e^{i\mathbf{k}\cdot\mathbf{r}} + r^{-1}\, e^{ikr} f(\theta, \phi). \tag{6.4.14}$$

The differential scattering cross-section is accordingly (cf. 5.10.7)

$$\sigma(\theta, \phi) = |f(\theta, \phi)|^2 = \left|\frac{m}{2\pi\hbar^2}\int e^{-i\mathbf{k}'\cdot\mathbf{r}'} V(\mathbf{r}')\, \psi(\mathbf{r}')\, d^3\mathbf{r}'\right|^2. \tag{6.4.15}$$

Further progress requires an approximation to $\psi(\mathbf{r}')$ in the integrand. In the Born approximation, one simply puts in the original plane wave $e^{i\mathbf{k}\cdot\mathbf{r}}$. The integral then takes the form of a Fourier transform in the space of

$$\mathbf{K} \equiv \mathbf{k}' - \mathbf{k}. \tag{6.4.16}$$

The quantity $\hbar\mathbf{K}$ corresponds to the *momentum transfer* in a scattering event. In terms of

† The second-order term gives a factor of order $e^{ikr'^2/r}$, which is close to unity when $r \gg kr'^2$.

$$V(\mathbf{K}) = (2\pi)^{-3} \int V(\mathbf{r})\, e^{-i\mathbf{K}\cdot\mathbf{r}}\, d^3\mathbf{r} \qquad (6.4.17)$$

we obtain a particularly simple expression for the cross-section:

$$\sigma(\mathbf{K}) = \left| \frac{4\pi^2 m}{\hbar^2} V(\mathbf{K}) \right|^2. \qquad (6.4.18)$$

The validity of the Born approximation rests on the dominance of the unscattered part of the wavefunction, i.e.,

$$|e^{i\mathbf{k}\cdot\mathbf{r}}|^2 \gg |r^{-1} e^{ikr} f(\theta, \phi)|^2. \qquad (6.4.19)$$

Integrating (6.4.19) over a sphere of radius a, measuring the effective range of the potential, we obtain approximately

$$a^2 \gg \sigma \qquad (6.4.20)$$

where σ is the total cross-section. Since a^2 approximates the geometrical cross-section of the target, the Born treatment evidently applies to scattering processes of relatively low probability. By writing

$$V(\mathbf{K}) = (2\pi)^{-3} V_0 a^3 \qquad (6.4.21)$$

in which V_0 is an effective average potential (actually a function of K), the preceding inequality can be rearranged to

$$\frac{2\pi\hbar^2}{ma^2} \gg V_0. \qquad (6.4.22)$$

While this is a sufficient condition for the Born approximation, it is usually more restrictive than necessary. Sharper alternative criteria relating the incident wavenumber to the strength of the potential are given in textbooks.

For scattering by a central field $V(r)$, integration over angles in (6.4.17) leads to

$$\sigma(K) = \left| \frac{2m}{\hbar^2 K} \int_0^\infty r V(r) \sin Kr\, dr \right|^2. \qquad (6.4.23)$$

As a concrete example, we consider the Coulomb scattering of α-particles by atoms, as exemplified by Rutherford's classic experiments. The repulsive interaction with the atomic nucleus corresponds to a potential energy

$$V(r) = \frac{ZZ'e^2}{r} \qquad (6.4.24)$$

in which Z and Z' $(=2)$ are, respectively, the atomic numbers of the target nucleus and the α-particle. However (6.4.24) in (6.4.23) results in a divergent integral, indicating that the pure Coulomb field is not sufficiently "short range" for the simple Born approximation to apply. In actual fact, the nuclear charge is partially "screened" by the atomic electrons. A screened Coulomb field can be simply approximated by a potential

$$V(r) = \frac{ZZ'e^2}{r} e^{-\alpha r}. \tag{6.4.25}$$

Calculations according to the Thomas–Fermi statistical model give α of the order of $Z^{\frac{1}{3}}/a_0$. With (6.4.25) in (6.4.23) and the definite integral

$$\int_0^\infty \sin Kr\, e^{-\alpha r}\, dr = \frac{K}{K^2 + \alpha^2} \tag{6.4.26}$$

we obtain

$$\sigma(K) = \left[\frac{2mZZ'e^2}{\hbar^2(K^2 + \alpha^2)}\right]^2. \tag{6.4.27}$$

Since, for elastic scattering $k = k'$, we find

$$K = 2k \sin\frac{\theta}{2} \tag{6.4.28}$$

where θ is the scattering angle. Under typical conditions, the momentum transfer is sufficiently large that α^2 can be neglected in the denominator of (6.4.27). If now we introduce the energy (6.4.3), we obtain Rutherford's famous scattering formula

$$\sigma(\theta) = \left(\frac{ZZ'e^2}{4E}\right)^2 \operatorname{cosec}^4\frac{\theta}{2}. \tag{6.4.29}$$

The quantum constant $\hbar$ does not occur and, indeed, the formula is in exact agreement with the result obtained by classical mechanics. Interestingly enough, it agrees as well with the *exact* quantum-mechanical calculation for the potential (6.4.24), another of those remarkable serendipities which seem to occur for the Coulomb field.

6.5 Green's Functions in Quantum Dynamics

In the preceding sections, Green's function techniques have been applied to the solution of quantum-mechanical perturbation problems formulated as inhomogeneous equations. But Green's functions are useful as well in fitting solutions of *homogeneous* equations to specified boundary conditions. This was shown explicitly in Section 6.1B for the case of Poisson's equation. Many problems of quantum dynamics belong to this latter category. We might, for example, require the solution of the time-dependent Schrödinger equation

$$\left(i\hbar\frac{\partial}{\partial t} - \mathcal{H}\right)\Psi(q, t) = 0 \tag{6.5.1}$$

consistent with an *initial condition* $\Psi(q, t_0)$. The Green's function for the Schrödinger operator is determined by

$$\left(i\hbar\frac{\partial}{\partial t} - \mathcal{H}\right)G(qt, q't') = \delta(q - q')\delta(t - t'). \tag{6.5.2}$$

The appropriate causal Green's function $G^{(+)}$ must vanish, as before, for $t' > t$:

$$G^{(+)}(qt, q't') = 0, \qquad t' > t\,. \tag{6.5.3}$$

The solution to (6.5.1) can be arrived at by a manipulation analogous to Green's theorem. First we write the complex conjugate of (6.5.2) for the *advanced* Green's function and the operator in primed coordinates:

$$\left[\left(i\hbar\frac{\partial}{\partial t'} - \mathcal{H}'\right)G^{(-)}(q't', qt)\right]^* = \delta(q' - q)\delta(t' - t)\,. \tag{6.5.4}$$

Also

$$\left(i\hbar\frac{\partial}{\partial t'} - \mathcal{H}'\right)\Psi(q', t') = 0\,. \tag{6.5.5}$$

Now multiply (6.5.4) by $\Psi(q', t')$, (6.5.5) by $G^{(-)}(q't', qt)^*$, subtract and integrate over q'. The terms containing $\mathcal{H}'$ cancel by virtue of the hermitian condition. There remains

$$-i\hbar\int\left[G^{(-)}(q't', qt)^*\frac{\partial}{\partial t'}\Psi(q', t') + \Psi(q', t')\frac{\partial}{\partial t'}G^{(-)}(q't', qt)^*\right]\mathrm{d}q'$$
$$= \Psi(q, t')\delta(t' - t). \tag{6.5.6}$$

Noting the conjugation relation (6.3.35) and that the bracket is a total time derivative, we have

$$-i\hbar\frac{\partial}{\partial t'}\int G^{(+)}(qt, q't')\Psi(q', t')\,\mathrm{d}q' = \Psi(q, t')\delta(t' - t). \tag{6.5.7}$$

Integrating over t' from some $t_0 < t$ to some $t_1 > t$ and recalling condition (6.5.3), we obtain

$$\Psi(q, t) = i\hbar\int G^{(+)}(qt, q't_0)\Psi(q', t_0)\,\mathrm{d}q' \tag{6.5.8}$$

which is the desired solution of the time-dependent Schrödinger equation.

Since the causal Green's function can be represented in the form (cf. 6.3.16)

$$G^{(+)}(qt, q't') = -\frac{i}{\hbar}\theta(t - t')K(qt, q't') \tag{6.5.9}$$

(6.5.8) becomes

$$\Psi(q, t) = \int K(qt, q't')\Psi(q', t')\,\mathrm{d}q' \tag{6.5.10}$$

having written t' in place of t_0. It is important to note that the integration runs only over the configuration variables but *not* the time. Time dependence is accounted for in a parametric fashion. Since K determines the time development of a quantum system, we shall call it the *dynamical Green's function.* The integral representation (6.5.10) is entirely equivalent to the evolution operator relation (4.9.1):

$$\Psi(q, t) = \mathscr{U}(t, t')\Psi(q, t'), \tag{6.5.11}$$

thus identifying $K(qt, q't')$ as the *integral kernal* of the evolution operator (cf 3.3.7). By the analog of (6.2.8),

$$\mathscr{U}(t, t')\delta(q - q') = K(qt, q't'). \tag{6.5.12}$$

Interpreted physically, this means that the dynamical Green's function represents the hypothetical state into which a perfectly localized wavepacket $\delta(q - q')$ evolves after a time $t - t'$. K must therefore be itself a solution of the time-dependent Schrödinger equation:

$$\left(i\hbar\frac{\partial}{\partial t} - \mathscr{H}\right)K(qt, q't') = 0 \tag{6.5.13}$$

corresponding to the initial condition

$$K(qt', q't') = \delta(q - q'). \tag{6.5.14}$$

The last identity also follows directly from (6.5.10) when $t = t'$.

The Green's function $K(qt, q't')$ plays a central role in Feynman's path-integral formulation of quantum mechanics.†

We shall specialize in what follows to conservative dynamical systems—Hamiltonian independent of time. In this case, dependence of K on the time variables occurs through $\tau = t - t'$ and we write $K = K(q, q', \tau)$. With the appropriate form of the evolution operator, eqn (6.5.12) becomes

$$\mathrm{e}^{-i\mathscr{H}\tau/\hbar}\delta(q - q') = K(q, q', \tau). \tag{6.5.15}$$

† R. P. Feynman, *Rev. Mod. Phys.* **20**, 367 (1948); R. P. Feynman and A. R. Hibbs, "Quantum Mechanics and Path Integrals", McGraw-Hill, New York, 1965.

By inversion

$$e^{i\mathscr{H}\tau/\hbar}K(q, q', \tau) = \delta(q - q') \tag{6.5.16}$$

This demonstrates, incidentally, that K is technically the Green's function for the *inverse* evolution operator. Moreover, since this operator is unitary, the formulas of Section 6.2A for unitary Green's functions pertain. Since $\mathscr{U}^{\dagger}(\tau) = \mathscr{U}(-\tau)$, the conjugation relation takes the form

$$K(q, q', \tau)^* = K(q', q, -\tau) \tag{6.5.17}$$

or, in terms of the original time variables,

$$K(qt, q't')^* = K(q't', qt)\,. \tag{6.5.18}$$

The spectral representation for the dynamical Green's function can be obtained by putting the closure relation for the eigenstates of $\mathscr{H}$ into (6.5.15):

$$K(q, q', \tau) = \mathop{\mathbf{S}}_{n} \psi_n^*(q')\psi_n(q)\, e^{-iE_n\tau/\hbar}. \tag{6.5.19}$$

This was previously obtained by Fourier transformation of $G(q, q', E)$ (cf 6.3.17). It is interesting to note that imaginary values of τ, such that

$$\tau = -i\hbar/kT, \tag{6.5.20}$$

transforms K into the canonical ensemble density matrix (or "temperature Green's function")

$$\rho(q, q') = \mathop{\mathbf{S}}_{n} \psi_n^*(q')\psi_n(q)\, e^{-E_n/kT}. \tag{6.5.21}$$

This explains why many of the same mathematical techniques are applicable to both field theory and statistical mechanics. In the limit $T \to 0$, (6.5.21) reduces to

$$\psi_0^*(q')\psi_0(q)\, e^{-E_0/kT} \tag{6.5.22}$$

in which $n = 0$ represents the ground state.

Introduction of time-dependent eigenfunctions

$$\Psi_n(q, t) = \psi_n(q)\, e^{-iE_nt/\hbar} \tag{6.5.23}$$

into the spectral representation (6.5.19) leads to the compact expression

$$K(qt, q't') = \mathop{\mathbf{S}}_{n} \Psi_n^*(q', t')\Psi_n(q, t). \tag{6.5.24}$$

A useful structural representation of the dynamical Green's function can be inferred from the solution of the corresponding Hamilton–Jacobi equation, in analogy with the argument of Section 4.2. Since, by (6.5.13), K is a solution of the time-dependent Schrödinger equation, we can write

$$K(qt, q't') = f(qt, q't') \exp\left[\frac{i}{\hbar} S(q, t)\right]. \tag{6.5.25}$$

In view of (6.5.18),

$$K(qt, q't') = K(q't', qt)^* = f(q't', qt)^* \exp\left[-\frac{i}{\hbar} S(q't')\right] \tag{6.5.26}$$

These imply in composite that

$$K(q, q', \tau) = F(q, q', \tau) \exp\left[\frac{i}{\hbar} S(q, q', \tau)\right] \tag{6.5.27}$$

such that S fulfills the exchange relation

$$S(q', q, -\tau) = -S(q, q', \tau). \tag{6.5.28}$$

Equation (6.5.27) also follows from the structure of the individual eigenfunctions (cf. 4.2.8):

$$\Psi_n(q, t) = F_n(q, t) \exp\left[\frac{i}{\hbar} S(q, t)\right]. \tag{6.5.29}$$

Thus, putting (6.5.29) into (6.5.24) gives (6.5.27) such that

$$F(q, q', \tau) = \mathop{\mathbf{S}}_{n} F_n^*(q', t')F_n(q, t). \tag{6.5.30}$$

In some simple cases discussed below, knowledge of Hamilton's principal function practically determines the Green's function. It remains to adjust the preexponential factor in conformity with the quantum-mechanical relations (6.5.13) and (6.5.14).

A. *Free particle*

The dynamical Green's function can be constructed using the spectral representation by virtue of the unique simplicity of free-particle eigenstates. For the one-dimensional free particle, putting

$$\psi_k(x) = (2\pi)^{-\frac{1}{2}} e^{ikx}, \qquad E(k) = \hbar^2k^2/2m \tag{6.5.31}$$

into (6.5.19), we obtain

$$K(x, x', \tau) = \frac{1}{2\pi}\int_{-\infty}^{\infty} e^{ik(x-x')} e^{-i\hbar k^2\tau/2m}\, dk\,. \tag{6.5.32}$$

The integral has the form of the Fourier transform of a gaussian. Application of (B.13) results in

$$K(x, x', \tau) = \left(\frac{m}{2\pi i\hbar\tau}\right)^{\frac{1}{2}} \exp\left[im(x - x')^2/2\hbar\tau\right]. \tag{6.5.33}$$

This follows alternatively from the Hamilton–Jacobi construction (6.5.27). Hamilton's principal function for the free particle is (cf. 1.4.19)

$$S(x, x', \tau) = \frac{m(x - x')^2}{2\tau} \tag{6.5.34}$$

so that

$$K(x, x', \tau) = F(x, x', \tau) \exp\left[im(x - x')^2/2\hbar\tau\right]. \tag{6.5.35}$$

In this instance, the normalization condition

$$K(x, x', 0) = \delta(x - x') \tag{6.5.36}$$

suffices to determine $F(x, x', \tau)$. Since (6.5.35) has the formal structure of a gaussian (cf.A.9), with

$$\sigma^2 = i\hbar\tau/m \tag{6.5.37}$$

the preexponential factor must be $(m/2\pi i\hbar\tau)^{\frac{1}{2}}$, in agreement with (6.5.33).

Given this Green's function, we can attack any initial-value problem in free-particle dynamics. Knowing the state at time $t = 0$, we can write

$$\Psi(x, t) = \int_{-\infty}^{\infty} K(x, x', t)\, \Psi(x', 0)\, \mathrm{d}x'. \tag{6.5.38}$$

We can, for example, obtain in a single step the evolution of a gaussian wave-packet, considered in Section 5.6. The initial state of a packet with $\langle x \rangle = x_0$, $\Delta x = \sigma_0$, $\langle p \rangle = \hbar k_0$ has the form (cf. 5.5.15)

$$\Psi(x', 0) = (2\pi\sigma_0^2)^{\frac{1}{4}}\, \mathrm{e}^{-(x' - x_0)^2/4\sigma_0^2}\, \mathrm{e}^{ik(x' - x_0)}. \tag{6.5.39}$$

Putting (6.5.33) and (6.5.39) into (6.5.38),

$$\Psi(x, t) = (m/2\pi i\hbar t)^{\frac{1}{2}}(2\pi\sigma_0^2)^{-\frac{1}{4}} \int_{-\infty}^{\infty} \exp\left[-\frac{(x' - x_0)^2}{4\sigma_0^2} + \frac{im(x - x')^2}{2\hbar t} + ik(x' - x_0)\right] \mathrm{d}x'. \tag{6.5.40}$$

Integration is straightforward, although a bit tedious, resulting in

$$\Psi(x, t) = (2\pi\Sigma^2)^{-\frac{1}{4}} \exp\left[\frac{im(x - x_0)^2}{2\hbar t_2}\right] \exp\left[-\frac{im\sigma_0(x - x_0 - \hbar k_0 t/m)^2}{2\hbar\sigma_0\Sigma}\right] \tag{6.5.41}$$

where (cf. 5.6.6)

$$\Sigma(t) = \sigma_0\left(1 + \frac{i\hbar t}{2m\sigma_0^2}\right). \tag{6.5.42}$$

It is not obvious that (6.5.41) is equal to (5.6.7), but evaluation of the probability density gives the identical result

$$\rho(x, t) = (2\pi\Sigma^2)^{-\frac{1}{2}} \exp\left[-(x - x_0 - \hbar k_0 t/m)^2/2\Sigma^2\right]. \tag{6.5.43}$$

By analogous arguments, we can determine the Green's function for the

3-dimensional free particle:

$$K(\mathbf{r}, \mathbf{r}', \tau) = \left(\frac{m}{2\pi i\hbar\tau}\right)^{\frac{3}{2}} \exp\left[im(\mathbf{r} - \mathbf{r}')^2/2\hbar\tau\right]. \tag{6.5.44}$$

It is instructive to verify that this can be obtained as well by Fourier transformation of the energy Green's function. We have seen that (cf. 6.5.9)

$$G^{(+)}(\mathbf{r}, \mathbf{r}', \tau) = -\frac{i}{\hbar}\theta(\tau)K(\mathbf{r}, \mathbf{r}', \tau). \tag{6.5.45}$$

Also (cf. 6.3.25)

$$G^{(-)}(\mathbf{r}, \mathbf{r}', \tau) = \frac{i}{\hbar}\theta(-\tau)K(\mathbf{r}, \mathbf{r}', \tau) \tag{6.5.46}$$

so that

$$K(\mathbf{r}, \mathbf{r}', \tau) = i\hbar\left[G^{(+)}(\mathbf{r}, \mathbf{r}', \tau) - G^{(-)}(\mathbf{r}, \mathbf{r}', \tau)\right]. \tag{6.5.47}$$

According to (6.3.10)

$$G^{(\pm)}(\mathbf{r}, \mathbf{r}', \tau) = \frac{1}{2\pi\hbar}\int_{-\infty}^{\infty} G^{(\pm)}(\mathbf{r}, \mathbf{r}', E)\, e^{-iE\tau/\hbar}\, dE \tag{6.5.48}$$

so that

$$K(\mathbf{r}, \mathbf{r}', \tau) = \frac{i}{2\pi}\int_{-\infty}^{\infty} \left[G^{(+)}(\mathbf{r}, \mathbf{r}', E) - G^{(-)}(\mathbf{r}, \mathbf{r}', E)\right] e^{-iE\tau/\hbar}\, dE. \tag{6.5.49}$$

The free-particle Green's functions for $E > 0$ are given by (6.4.6):

$$G^{(\pm)}(R, E) = -\frac{m}{2\pi\hbar^2}\frac{e^{\pm ikR}}{R} \tag{6.5.50}$$

in terms of $R = |\mathbf{r} - \mathbf{r}'|$ and $k = (2mE)^{\frac{1}{2}}/\hbar$. We require as well the Green's function for $E < 0$, in order to complete the integration in (6.5.49). Writing

$$E = -h^2\kappa^2/2m \tag{6.5.51}$$

this Green's function is the solution of

$$\frac{\hbar^2}{2m}(\nabla^2 - \kappa^2)G(R, \kappa) = \delta(\mathbf{R}). \tag{6.5.52}$$

By the method of Section 6.1A, we obtain

$$G(R, E < 0) = -\frac{m}{2\pi\hbar^2}\frac{e^{-\kappa R}}{R} \tag{6.5.53}$$

which, incidentally, has the form of the Yukowa potential. The negative-E

portion of the integral (6.5.49) is now seen to drop out, since $G(R, E < 0)$ has the same form for both $E \pm i0^+$. Thus, using (6.5.50) and changing the variable of integration to k, we obtain

$$K(R, \tau) = -\frac{i}{4\pi^2 R}\int_0^\infty (e^{ikR} - e^{-ikR})\, e^{-i\hbar k^2\tau/2m} k\, dk$$

$$= -\frac{i}{4\pi^2 R}\frac{\partial}{\partial R}\int_{-\infty}^{\infty} e^{ikR}\, e^{-i\hbar k^2\tau/2m}\, dk\,. \tag{6.5.54}$$

The last integral represents the inverse transform to (B.13). We find

$$\frac{1}{\sqrt{(2\pi)}}\int_{-\infty}^{\infty} e^{ikR}\, e^{-i\hbar k^2\tau/2m} = \left(\frac{m}{i\hbar\tau}\right)^{\frac{1}{2}} e^{-mR^2/2i\hbar\tau} \tag{6.5.55}$$

which leads to

$$K(R, \tau) = \left(\frac{m}{2\pi i\hbar\tau}\right)^{\frac{3}{2}} e^{imR^2/2\hbar\tau} \tag{6.5.56}$$

in agreement with (6.5.44).

B. *Linear Harmonic Oscillator*

The Hamiltonian for this system can be written

$$\mathscr{H} = -\frac{\hbar^2}{2m}\frac{\partial^2}{\partial x^2} + \frac{m\omega^2}{2}x^2 \tag{6.5.57}$$

in terms of the oscillator frequency ω. Hamilton's principal function, worked out in Section 1.4 (cf. 1.4.29), can be expressed

$$S(x, x', \tau) = \frac{m\omega}{2}\left[(x^2 + x'^2)\cot\omega\tau - 2xx'\operatorname{cosec}\omega\tau\right]. \tag{6.5.58}$$

One can therefore apply the representation (6.5.27) to construction of the dynamical Green's function:

$$K(x, x', \tau) = F(x, x', \tau)\exp\left[\frac{i}{\hbar}S(x, x', \tau)\right]. \tag{6.5.59}$$

The preexponential function must now be determined such that

$$\left(i\hbar\frac{\partial}{\partial\tau} - \mathscr{H}\right)K(x, x', \tau) = 0 \tag{6.5.60}$$

and

$$K(x, x', 0) = \delta(x - x'). \tag{6.5.61}$$

Let us first try $F = F(\tau)$, independent of x and x', as in the free-particle case.

With (6.5.57) and (6.5.59) in (6.5.60), we obtain

$$i\hbar\frac{\mathrm{d}F}{\mathrm{d}\tau} - F\frac{\partial S}{\partial \tau} + \frac{\hbar^2}{2m}\left[\left(\frac{i}{\hbar}\right)^2 F\left(\frac{\partial S}{\partial x}\right)^2 + \frac{i}{\hbar}F\frac{\partial^2 S}{\partial x^2}\right] - \frac{m\omega^2}{2}x^2F = 0. \tag{6.5.62}$$

Now

$$\frac{\partial S}{\partial \tau} = -\frac{m\omega^2}{2}[(x^2 + x'^2)\operatorname{cosec}^2\omega\tau - 2xx'\operatorname{cosec}\omega\tau\operatorname{cotan}\omega\tau]$$

$$\frac{\partial S}{\partial x} = m\omega(x\operatorname{cotan}\omega\tau - x'\operatorname{cosec}\omega\tau)$$

$$\frac{\partial^2 S}{\partial x^2} = m\omega\operatorname{cotan}\omega\tau \tag{6.5.63}$$

Substitution into (6.5.62) leads to

$$\frac{\mathrm{d}F}{\mathrm{d}\tau} + \frac{\omega}{2}\operatorname{cotan}\omega\tau\, F = 0. \tag{6.5.64}$$

After separation of variables and integration:

$$\log F = -\tfrac{1}{2}\log|\sin\omega\tau| + \text{const} \tag{6.5.65}$$

or

$$F = \text{const}\,(\sin\omega\tau)^{-\frac{1}{2}}. \tag{6.5.66}$$

To fulfil the initial condition (6.5.61) we find

$$F(\tau) = \left(\frac{m\omega}{2\pi i\hbar\sin\omega\tau}\right)^{\frac{1}{2}} \tag{6.5.67}$$

noting that both $S(x, x', \tau)$ and $F(\tau)$ approach their free-particle counterparts as $\tau \to 0$. Written in full, the harmonic-oscillator Green's function is

$$K(x, x', \tau) = \left(\frac{m\omega}{2\pi i\hbar\sin\omega\tau}\right)^{\frac{1}{2}}\exp\left\{\frac{im\omega}{2\hbar\sin\omega\tau} \times [(x^2 + x'^2)\cos\omega\tau - 2xx']\right\} \tag{6.5.68}$$

having reverted to the sine and cosine in the exponential factor.

According to the spectral representation (6.5.19), knowledge of the Green's function makes accessible, in principle, everything about the eigenstates of a system. To determine the eigenvalue spectrum, it is convenient to integrate out dependence on x and x'. The most obvious way to do this is to compute the trace:

$$\operatorname{Tr} K(x, x', \tau) \equiv \int_{-\infty}^{\infty} K(x, x', \tau)\, dx. \tag{6.5.69}$$

In terms of the spectral representation

$$\operatorname{Tr} K = \mathop{\mathsf{S}}_{n}(\psi_n, \psi_n)\, e^{-iE_n\tau/\hbar}. \tag{6.5.70}$$

If, however, the spectrum contains *any* continuum portion, we should have some $(\psi_\nu, \psi_\nu) = \infty$ which would cause Tr K to diverge. If, on the other hand, the spectrum were entirely discrete, then each $(\psi_n, \psi_n) = 1$ and

$$\operatorname{Tr} K = \sum_n e^{-iE_n\tau/\hbar} \qquad \text{(discrete spectrum)}. \tag{6.5.71}$$

The trace will therefore exist if the summation converges. We arrive thereby at (at least) a necessary condition for a discrete eigenvalue spectrum: that the trace of the Green's function exist. Once Tr K is determined, the eigenvalue spectrum follows easily by Fourier expansion or, more formally, by inversion to†

$$g(\omega) = \frac{1}{2\pi}\int_{-\infty}^{\infty} \operatorname{Tr} K\, e^{i\omega\tau}\, d\tau = \sum_n \delta(\omega - \omega_n) \qquad (\omega_n = E_n/\hbar). \tag{6.5.72}$$

To apply this spectral analysis to the harmonic-oscillator Green's function, we find, using (6.5.68),

$$\begin{aligned}\operatorname{Tr} K &= \left(\frac{m\omega}{2\pi i\hbar \sin\omega\tau}\right)^{\frac{1}{2}} \int_{-\infty}^{\infty} \exp\left\{-\frac{im\omega(1-\cos\omega\tau)}{\hbar\sin\omega\tau}x^2\right\} dx \\ &= \left(\frac{m\omega}{2\pi i\hbar \sin\omega\tau}\right)^{\frac{1}{2}} \left[\frac{\pi\hbar\sin\omega\tau}{im\omega(1-\cos\omega\tau)}\right]^{\frac{1}{2}} \\ &= [-2(1-\cos\omega\tau)]^{-\frac{1}{2}} = (2i\sin\tfrac{1}{2}\omega\tau)^{-1}.\end{aligned} \tag{6.5.73}$$

The fact that Tr K does exist (except at $\tau = 0, \pm 2\pi/\omega$, etc.) indicates that the harmonic-oscillator eigenvalue spectrum is purely discrete. Expanding Tr K in a complex Fourier series gives

$$\begin{aligned}\operatorname{Tr} K &= (e^{i\omega\tau/2} - e^{-i\omega\tau/2}) = e^{-i\omega\tau/2}(1 - e^{-i\omega\tau})^{-1} \\ &= \sum_{n=0}^{\infty} e^{-i(n+\frac{1}{2})\omega\tau}\end{aligned} \tag{6.5.74}$$

which identifies the eigenvalues

$$E_n = (n + \tfrac{1}{2})\hbar\omega, \qquad n = 0, 1, 2 \ldots \tag{6.5.75}$$

† On this principle, I have suggested a possible technique for determining eigenvalue spectra of quantum-mechanical systems: S. M. Blinder, *Int. J. Quant. Chem.* **1**, 285 (1967); *Theoret. chim. Acta* (*Berlin*) **24**, 382 (1972).

The eigenfunctions are a bit more difficult to obtain from the Green's function. The harmonic-oscillator eigenfunctions follow straightforwardly, however, if use is made of a summation formula for Hermite polynomials (due to Mehler):

$$\sum_{n=0}^{\infty} H_n(\xi)H_n(\xi')\frac{\theta^n}{2^n n!} = (1-\theta^2)^{-\frac{1}{2}} \exp\left[\frac{2\xi\xi'\theta - (\xi^2+\xi'^2)\theta^2}{1-\theta^2}\right]. \tag{6.5.76}$$

By means of the substitutions

$$\theta = \mathrm{e}^{-i\omega\tau} \tag{6.5.77}$$

$$2i \sin \omega\tau = \frac{1-\theta^2}{\theta}, \qquad 2\cos\omega\tau = \frac{1+\theta^2}{\theta} \tag{6.5.78}$$

$$\beta = m\omega/\hbar \tag{6.5.79}$$

and

$$\xi = \beta^{\frac{1}{2}}x, \qquad \xi' = \beta^{\frac{1}{2}}x' \tag{6.5.80}$$

the Green's function (6.5.68) can be rearranged to

$$K = \left[\frac{\beta\theta}{\pi(1-\theta^2)}\right]^{\frac{1}{2}} \exp\left[\frac{2\xi\xi'\theta - (\xi^2+\xi'^2)\theta^2}{1-\theta^2} - \frac{\xi^2+\xi'^2}{2}\right]. \tag{6.5.81}$$

Thus

$$K = \left(\frac{\beta}{\pi}\right)^{\frac{1}{2}} \sum_{n=0}^{\infty} H_n(\xi)H_n(\xi')\,\mathrm{e}^{-(\xi^2+\xi'^2)/2}\frac{\theta^{n+\frac{1}{2}}}{2^n n!} \tag{6.5.82}$$

which identifies the normalized harmonic-oscillator eigenfunctions:

$$\psi_n(\xi) = \left(\frac{\beta}{\pi}\right)^{\frac{1}{4}} (2^n n!)^{-\frac{1}{2}} H_n(\xi)\,\mathrm{e}^{-\xi^2/2}. \tag{6.5.83}$$

Equation (6.5.81) can be interpreted as a temperature Green's function (cf. 6.5.20, 6.5.21) if

$$\theta = \mathrm{e}^{-\hbar\omega/kT}. \tag{6.5.84}$$

In the limit $T \to 0$ (or $\theta \to 0$), (6.5.81) reduces to

$$\left(\frac{\beta\theta}{\pi}\right)^{\frac{1}{2}} \mathrm{e}^{-(\xi^2+\xi'^2)/2} \tag{6.5.85}$$

which, according to (6.5.22), identifies

$$\psi_0(\xi) = \left(\frac{\beta}{\pi}\right)^{\frac{1}{4}} \mathrm{e}^{-\xi^2/2}, \qquad E_0 = \tfrac{1}{2}\hbar\omega. \tag{6.5.86}$$

For the 3-dimensional isotropic harmonic oscillator, the dynamical Green's function is given by

$$K(\mathbf{r}, \mathbf{r}', \tau) = \left(\frac{m\omega}{2\pi i\hbar \sin \omega\tau}\right)^{\frac{3}{2}} \exp\left\{\frac{im\omega}{2\hbar \sin \omega\tau} \times [(r^2 + r'^2)\cos \omega\tau - \mathbf{r}\cdot\mathbf{r}']\right\} \quad (6.5.87)$$

which is the product of identical factors for x, y and z motions. To determine the eigenvalue spectrum, we evaluate

$$\mathrm{Tr}\, K = \int K(\mathbf{r}, \mathbf{r}', \tau)\, \mathrm{d}^3\mathbf{r} = (2i \sin \tfrac{1}{2}\omega\tau)^{-3}. \quad (6.5.88)$$

This has the Fourier expansion

$$\mathrm{Tr}\, K = \sum_{n=0}^{\infty} \frac{(n+1)(n+2)}{2} \mathrm{e}^{-i(n+\frac{3}{2})\omega\tau} \quad (6.5.89)$$

which identifies the energy eigenvalues

$$E_n = (n + \tfrac{3}{2})\hbar\omega, \qquad n = 0, 1, 2 \ldots \quad (6.5.90)$$

with degeneracies

$$g_n = \frac{(n+1)(n+2)}{2}. \quad (6.5.91)$$

7

Theory of Transitions

7.1 Transition Probability

Let an arbitrary state $\Psi(q, t)$ of a nonconservative system $[\mathscr{H} = \mathscr{H}(t)]$ be represented as a superposition of eigenstates of some observable $\mathscr{A}$ (cf. 4.11.6):

$$\Psi(q, t) = \mathop{\mathbf{S}}_{n} c_n(t)\, \phi_n(q) \tag{7.1.1}$$

in which the basis $\{\phi_n(q)\}$ is assumed complete and orthonormal:

$$(\phi_n, \phi_{n'}) = \delta_{nn'} \tag{7.1.2}$$

Suppose now that, at some initial time t_0, the system is placed in one of these eigenstates, say,

$$\Psi(q, t_0) = \phi_m(q) \tag{7.1.3}$$

corresponding to

$$c_n(t_0) = \delta_{nm}\,. \tag{7.1.4}$$

As the system evolves in time thereafter, its wavefunction $\Psi(q, t)$ will remain normalized, its expansion coefficients satisfying (cf. 3.9.21)

$$\mathop{\mathbf{S}}_{n} |c_n(t)|^2 = 1\,. \tag{7.1.5}$$

In accordance with the discussion in Section 3.9A, one can interpret each $|c_n(t)|^2$ as the probability that the system will be observed at the later time t in the eigenstate $\phi_n(q)$. This corresponds to a *transition probability*

$$P(m, t_0 \rightarrow n, t) = |c_n(t)|^2 \tag{7.1.6}$$

also to be designated $P_{m\rightarrow n}$ for brevity. For a member of a continuous spectrum, say ν, $|c_\nu(t)|^2$ represents a *transition probability density* per quantum state:

$$\rho_{\mu\rightarrow\nu} = |c_\nu(t)|^2. \tag{7.1.7}$$

The state labelled μ can be of either discrete or continuum character. For transitions to some *group* of continuum states $\{\nu\}$ one can again define a discrete transition probability

$$P_{\mu\to\{\nu\}} = \int_{\{\nu\}} \rho_{\mu\to\nu}\, d\nu\,. \tag{7.1.8}$$

One could represent the above physical situation in terms of the evolution operator as follows:

$$\Psi(q,t) = \mathscr{U}(t,t_0)\,\phi_m(q). \tag{7.1.9}$$

Substituting the expansion (7.1.1) for $\Psi(q,t)$, taking the inner product with $\phi_n(q)$, and using (7.1.2), we find

$$c_n(t) = \langle n|\mathscr{U}(t,t_0)|m\rangle. \tag{7.1.10}$$

The transition probability can therefore be represented†

$$P(m,t_0 \to n,t) = |\langle n|\mathscr{U}(t,t_0)|m\rangle|^2. \tag{7.1.11}$$

This result can be generalized further to give the probability that the system originally in the eigenstate $\phi_m(q)$ of $\mathscr{A}$ at time t_0 makes a transition by time t to the eigenstate $\chi_n(q)$ of a *different* observable $\mathscr{B}$. Denoting the respective eigenkets by $|a_m\rangle$ and $|b_n\rangle$, we deduce the transition probability

$$P(a_m,t_0 \to b_n,t) = |\langle b_n|\mathscr{U}(t,t_0)|a_m\rangle|^2. \tag{7.1.12}$$

The matrix elements $(\chi_n, \mathscr{U}(t,t_0)\phi_m)$ are, incidentally, known as *transition amplitudes*. In terms of the dynamical Green's function (cf. Section 6.5)

$$(\chi_n, \mathscr{U}(t,t_0)\phi_m) = \iint \chi_n^*(q)\,K(qt,q't_0)\,\phi_m(q')\,d\tau\,d\tau'. \tag{7.1.13}$$

Accordingly, $K(qt,q't')$ is also called a *transition kernel* in Feynman's path-integral formulation of quantum mechanics. In Heisenberg picture, where the initial state is represented by $\Phi_m^H(q,t_0) = \phi_m^S(q)$ (cf. 4.14.11) and the final state by $X_n^H(q,t) = \mathscr{U}^\dagger(t,t_0)\chi_n^S(q)$ (cf. 4.14.8), the transition amplitude becomes simply the overlap integral $(X_n^H(q,t), \Phi_m^H(q,t))$.

† To exclude the "nontransition" $m \to m$, one sometimes defines the *transition matrix*

$$\mathscr{T} \equiv \mathscr{U} - 1.$$

Equation (7.1.11) can accordingly be written

$$P_{m\to n} = |\langle n|\mathscr{T}|m\rangle|^2,$$

for $n \neq m$.

The evolution operator for $t = \infty$, $t_0 = -\infty$ is the *S-matrix* (scattering matrix):

$$\mathscr{S} \equiv \mathscr{U}(\infty, -\infty).$$

S-matrix formalism, introduced by Heisenberg, is particularly appropriate for describing complicated scattering events entirely in terms of observable initial and final states.

Transforming a transition amplitude using (3.3.4),

$$\langle b_n|\mathscr{U}(t, t_0)|a_m\rangle = \langle a_m|\mathscr{U}^\dagger(t, t_0)|b_n\rangle^*. \tag{7.1.14}$$

Noting (cf. 4.9.4)

$$\mathscr{U}^\dagger(t, t_0) = \mathscr{U}(t_0, t), \tag{7.1.15}$$

it follows that

$$P_{m\to n} = P_{n\to m}. \tag{7.1.16}$$

Thus the probability of a given transition is exactly equal to that of the reverse transition. This is an instance of the *principle of detailed balancing*.

Summing (7.1.11) or (7.1.12) over n, we find

$$\underset{n}{\mathbf{S}}\, P_{m\to n} = \underset{n}{\mathbf{S}}\, \langle n|\mathscr{U}|m\rangle^* \langle n|\mathscr{U}|m\rangle = \underset{n}{\mathbf{S}}\, \langle m|\mathscr{U}^\dagger|n\rangle \langle n|\mathscr{U}|m\rangle$$

$$= \langle m|\mathscr{U}^\dagger\mathscr{U}|m\rangle = \langle m|m\rangle = 1, \tag{7.1.17}$$

making use of matrix multiplication and the unitarity of $\mathscr{U}$. Summing over m gives an analogous result. Therefore, the sum of all possible transitions *from* or *to* a given state results in

$$\underset{n}{\mathbf{S}}\, P_{m\to n} = \underset{m}{\mathbf{S}}\, P_{m\to n} = 1. \tag{7.1.18}$$

Since each term is positive definite, we have necessarily

$$0 \leqslant P_{m\to n} \leqslant 1. \tag{7.1.19}$$

Any apparent violation of (7.1.19) must arise from a defect of approximation —a truncated perturbation expansion, for example.

7.2 Interaction Picture

For a conservative system with Hamiltonian $\mathscr{H}_0$, the equation of motion in Schrödinger picture is

$$i\hbar\, \partial\Psi^S(q, t)/\partial t = \mathscr{H}_0\Psi^S(q, t). \tag{7.2.1}$$

The formal solution is given by

$$\Psi^S(q, t) = \mathscr{U}_0(t, t_0)\,\Psi^S(q, t_0) \tag{7.2.2}$$

with

$$\mathscr{U}_0(t, t_0) = \exp\left[-\frac{i}{\hbar}(t - t_0)\mathscr{H}_0\right], \tag{7.2.3}$$

the evolution operator appropriate to a time-independent Hamiltonian (cf. 4.8.16). The evolution operator itself obeys the equation of motion

$$i\hbar \frac{\partial \mathscr{U}_0}{\partial t} = \mathscr{H}_0 \mathscr{U}_0. \tag{7.2.4}$$

Suppose now that the above system is perturbed by a small time-dependent potential $\mathscr{V}(t)$.

The resulting nonconservative system described by

$$\mathscr{H}(t) = \mathscr{H}_0 + \mathscr{V}(t) \tag{7.2.5}$$

follows the equation of motion

$$i\hbar\, \partial \Psi^S(q,t)/\partial t = [\mathscr{H}_0 + \mathscr{V}(t)]\, \Psi^S(q,t). \tag{7.2.6}$$

If $\mathscr{V}(t)$ is relatively small, however, the dynamical behaviour of the perturbed system will be preponderantly determined by $\mathscr{H}_0$. To isolate the residual influence of the perturbing potential, we define a state function $\Psi^{\mathrm{I}}(q,t)$ such that

$$\Psi^S(q,t) = \mathscr{U}_0(t,t_0)\, \Psi^{\mathrm{I}}(q,t). \tag{7.2.7}$$

If $\mathscr{V}(t)$ were zero, (7.2.7) would reduce simply to (7.2.2). In the general case, the unitary transformation represented by (7.2.7) establishes the *interaction picture.*

Substituting (7.2.7) into (7.2.6), we have

$$i\hbar \frac{\partial \mathscr{U}_0}{\partial t} \Psi^{\mathrm{I}} + i\hbar \mathscr{U}_0 \frac{\partial \Psi^{\mathrm{I}}}{\partial t} = \mathscr{H}_0 \mathscr{U}_0 \Psi^{\mathrm{I}} + \mathscr{V}(t) \mathscr{U}_0 \Psi^{\mathrm{I}}. \tag{7.2.8}$$

The first terms on each side cancel, by virtue of (7.2.4). The dynamical influence of $\mathscr{H}_0$ is thereby subtracted out. Premultiplying by $\mathscr{U}_0^\dagger$ and writing

$$\mathscr{V}^{\mathrm{I}}(t) = \mathscr{U}_0^\dagger(t,t_0)\mathscr{V}(t)\mathscr{U}_0(t,t_0), \tag{7.2.9}$$

we obtain

$$i\hbar \frac{\partial \Psi^{\mathrm{I}}}{\partial t} = \mathscr{V}^{\mathrm{I}} \Psi^{\mathrm{I}}, \tag{7.2.10}$$

which represents the interaction-picture transform of (7.2.6).

An arbitrary dynamical variable A is represented in the interaction picture by the operator

$$\mathscr{A}^{\mathrm{I}}(t) = \mathscr{U}_0^\dagger(t,t_0)\mathscr{A}^S \mathscr{U}_0(t,t_0). \tag{7.2.11}$$

$\mathscr{H}_0$ is itself the same in both pictures. In analogy with (4.14.19), the equation of motion for $\mathscr{A}$ is given by

$$\frac{\mathrm{d}\mathscr{A}^{\mathrm{I}}}{\mathrm{d}t} = \frac{\partial \mathscr{A}^{\mathrm{I}}}{\partial t} + (i\hbar)^{-1}[\mathscr{A}^{\mathrm{I}}, \mathscr{H}_0]. \tag{7.2.12}$$

Equations (7.2.10) and (7.2.12) show that, in the interaction picture, *both* state vectors *and* operators are moving in Hilbert space. Their respective time developments respond, however, to different portions of the total Hamiltonian: Ψ^{I} to $\mathscr{V}(t)$, $\mathscr{A}^{\mathrm{I}}$ to $\mathscr{H}_0$. The interaction picture is hence intermediate between Schrödinger and Heisenberg pictures, reducing to the latter as $\mathscr{V}(t)$ approaches zero.

When the Hamiltonian for a system contains a constant term E_0—for example, relativistic rest energy m_0c^2—all Schrödinger wave functions carry the phase factors $\mathrm{e}^{-iE_0t/\hbar}$. These common factors can, however, be eliminated by transforming to interaction picture, using

$$\mathscr{U}_0 = \mathrm{e}^{-iE_0t/\hbar}. \tag{7.2.13}$$

(Heisenberg picture shares this advantage since $\mathscr{U}_0$ is a factor of the overall transformation $\mathscr{U}$.) At the same time, dynamical variables are invariant since (7.2.13) commutes through $\mathscr{A}^S$ in (7.2.11).

The preceding result has an implication of deep significance. Delineation of a physical system focuses upon some specific portion of the Universe, the remainder serving, in effect, as an unperturbed background. By disregarding the rest of the Universe in any theoretical discussion we are thus implicitly utilizing an interaction picture.

To obtain the general solution of the interaction picture Schrödinger equation, one can proceed in complete analogy with the calculation of Section 4.8, with $\mathscr{V}^{\mathrm{I}}$ replacing $\mathscr{H}$. The solution can be represented as

$$\Psi^{\mathrm{I}}(q, t) = \mathscr{U}_{\mathrm{I}}(t, t_0)\,\Psi^{\mathrm{I}}(q, t_0). \tag{7.2.14}$$

In analogy with (4.8.6), the time-integral expansion for the evolution operator is given by

$$\mathscr{U}_{\mathrm{I}}(t, t_0) = 1 + \sum_{n=1}^{\infty} \mathscr{U}_{\mathrm{I}}^{(n)}(t, t_0)$$

where

$$\mathscr{U}_{\mathrm{I}}^{(n)}(t, t_0) \equiv (i\hbar)^{-n} \int_{t_0}^{t}\int_{t_0}^{t'} \cdots \int_{t_0}^{t'} \mathscr{V}^{\mathrm{I}}(t')\mathscr{V}^{\mathrm{I}}(t'')\ldots\mathscr{V}^{\mathrm{I}}(t^{(n)})\,\mathrm{d}t'\,\mathrm{d}t''\ldots\mathrm{d}t^{(n)}. \tag{7.2.15}$$

More compactly, in analogy with (4.8.15),

$$\mathscr{U}_{\mathrm{I}}(t, t_0) = \mathscr{P}\exp\left[-\frac{i}{\hbar}\int_{t_0}^{t} \mathscr{V}^{\mathrm{I}}(t')\,\mathrm{d}t'\right]. \tag{7.2.16}$$

The original equation (7.2.6) in Schrödinger picture has the formal solution

$$\Psi^S(q, t) = \mathscr{U}(t, t_0)\,\Psi^S(q, t_0). \tag{7.2.17}$$

Making use of (7.2.7) and (7.2.14), noting that $\Psi^S(q, t_0) = \Psi^I(q, t_0)$, this evolution operator is shown to have the composite structure†

$$\mathscr{U}(t, t_0) = \mathscr{U}_0(t, t_0)\,\mathscr{U}_I(t, t_0). \tag{7.2.18}$$

More explicitly,

$$\mathscr{U}(t, t_0) = \exp\left[-\frac{i}{\hbar}(t - t_0)\mathscr{H}_0\right]\mathscr{P}\exp\left[-\frac{i}{\hbar}\int_{t_0}^{t}\mathscr{V}^I(t')\,dt'\right]. \tag{7.2.19}$$

The identity (cf. 4.9.9)

$$\mathscr{U}(t, t_0) = \mathscr{U}^\dagger(t_0 t) \tag{7.2.20}$$

implies that

$$\mathscr{U}(t, t_0) = \mathscr{U}_I(t, t_0)\,\mathscr{U}_0(t, t_0) \tag{7.2.21}$$

showing that $\mathscr{U}_0$ and $\mathscr{U}_I$ commute (even when $\mathscr{H}_0$ and $\mathscr{V}(t)$ do not). Using (cf. 4.9.5)

$$\mathscr{U}(t, t_0) = \mathscr{U}(t, 0)\,\mathscr{U}(0, t_0) \tag{7.2.22}$$

we can also write

$$\mathscr{U}(t, t_0) = \mathscr{U}_0(t, 0)\,\mathscr{U}_I(t, 0)\,\mathscr{U}_I(0, t_0)\,\mathscr{U}_0(0, t_0) \tag{7.2.23}$$

or

$$\mathscr{U}(t, t_0) = e^{-i\mathscr{H}_0 t/\hbar}\,\mathscr{U}_I(t, t_0)\,e^{i\mathscr{H}_0 t_0/\hbar}. \tag{7.2.24}$$

It also follows directly from (7.2.6) that

$$\mathscr{U}(t, t_0) = \mathscr{P}\exp\left[-\frac{i}{\hbar}\int_{t_0}^{t}\{\mathscr{H}_0 + \mathscr{V}(t')\}\,dt'\right]. \tag{7.2.25}$$

As a final note, recall the Fourier representation of time-dependent wave functions discussed in Section 4.11. Applied to the interaction-picture Schrödinger equation (7.2.10), the analog of (4.11.6) can be written

$$\Psi^I(q, t) = \mathop{\mathbf{S}}_{n} C_n(t)\psi_n(q). \tag{7.2.26}$$

† This can be demonstrated alternatively from the evolution operator equations of motion

$$i\hbar\frac{\partial\mathscr{U}_0}{\partial t} = \mathscr{H}_0\mathscr{U}_0, \qquad i\hbar\frac{\partial\mathscr{U}_I}{\partial t} = \mathscr{V}^I\mathscr{U}_I = \mathscr{U}_0^\dagger\mathscr{V}\mathscr{U}_0\mathscr{U}_I.$$

Postmultiplying the first equation by $\mathscr{U}_I$, premultiplying the second by $\mathscr{U}_0$ and adding, we obtain

$$i\hbar\frac{\partial}{\partial t}(\mathscr{U}_0\mathscr{U}_I) = (\mathscr{H}_0 + \mathscr{V})\mathscr{U}_0\mathscr{U}_I$$

which must represent the solution to

$$i\hbar\frac{\partial\mathscr{U}}{\partial t} = (\mathscr{H}_0 + \mathscr{V})\mathscr{U}.$$

The most appropriate basis functions are $\{\psi_n(q)\}$, the eigenfunctions of $\mathscr{H}_0$. By virtue of (7.2.7) (taking $t_0 = 0$ for simplicity), (7.2.26) is equivalent to the Schrödinger wavefunction

$$\Psi^S(q, t) = \mathscr{U}_0(t, 0)\,\Psi^I(q, t) = \mathop{\mathbf{S}}_n C_n(t)\,\mathscr{U}_0(t, 0)\,\psi_n(q). \tag{7.2.27}$$

Noting that

$$\mathscr{U}_0(t, 0)\,\psi_n(q) = e^{-i\omega_n t}\psi_n(q) \tag{7.2.28}$$

we find

$$\Psi^S(q, t) = \mathop{\mathbf{S}}_n C_n(t)\,e^{-i\omega_n t}\psi_n(q). \tag{7.2.29}$$

Note the structural correspondence between (7.2.29) and the expansion (4.11.2) for a conservative system. In the present case, the coefficients $C_n(t)$ are time-dependent, reflecting the dynamical influence of the time-dependent perturbation $\mathscr{V}(t)$. The oscillatory factors $e^{-i\omega_n t}$ represent, on the other hand, the response to $\mathscr{H}_0$, the conservative part of the Hamiltonian. Expansion (1.2.29) provides a convenient starting point for the derivation of time-dependent perturbation theory in Section 7.4.

7.3 Time-dependent Perturbation Theory

Quantum-mechanical perturbation techniques are based on a partitioning of the Hamiltonian into two parts, one "simple", the other "small", viz.

$$\mathscr{H} = \mathscr{H}_0 + \mathscr{V}. \tag{7.3.1}$$

The first term, $\mathscr{H}_0$, represents some idealization of the system—either physical or mathematical. If $\mathscr{H}_0$ is time-independent (nearly always the case), it is assumed that the associated Schrödinger equation has been solved exactly (or nearly so) for all its eigenstates:

$$\mathscr{H}_0\psi_n(q) = \hbar\omega_n\psi_n(q). \tag{7.3.2}$$

The operator $\mathscr{V}$ represents a small correction or perturbation on the idealized system. It is "small", more precisely, in the sense that, for typical non-zero matrix elements,

$$|V_{mn}| \ll |H^0_{mn}|. \tag{7.3.3}$$

In stationary perturbation theory, $\mathscr{V}$ is also a time-independent operator and the object is to calculate the eigenvalues and eigenfunctions of $\mathscr{H}$. Time-dependent perturbation theory is concerned with problems in which $\mathscr{H}_0$ is perturbed by a time-varying field $\mathscr{V}(t)$. The theory was, in fact, developed specifically to treat the interaction of atomic and molecular systems with electromagnetic radiation.†

A time-dependent Hamiltonian has, of course, no eigenstates. Rather than considering modifications in the states of the system, one supposes that transitions occur among its unperturbed states. The object of time-dependent perturbation theory is accordingly to calculate transition probabilities.

Suppose the system described by

$$\mathscr{H}(t) = \mathscr{H}_0 + \mathscr{V}(t) \tag{7.3.4}$$

to be initially, at time t_0, in the eigenstate $\psi_m(q)$ of $\mathscr{H}_0$. The unperturbed system would, of course, remain in this eigenstate, exhibiting the characteristic oscillatory time dependence

$$\Psi_m(q, t) = e^{-i\omega_m t}\psi_m(q). \tag{7.3.5}$$

The perturbed system will, however, evolve into an admixture of eigenstates. This is most appropriately formulated in terms of the transition probabilities

$$P_{m\to n} = |\langle n|\mathscr{U}(t, t_0)|m\rangle|^2. \tag{7.3.6}$$

Interaction picture provides a convenient framework for time-dependent perturbation theory. For the Hamiltonian (7.3.4), the evolution operator can be put in the form (7.2.18). Therefore,

$$\begin{aligned}\langle n|\mathscr{U}(t, t_0)|m\rangle &= \langle n|\mathscr{U}_0(t, t_0)\,\mathscr{U}_{\mathrm{I}}(t, t_0)|m\rangle \\ &= e^{i\omega_n(t-t_0)}\langle n|\mathscr{U}_{\mathrm{I}}(t, t_0)|m\rangle,\end{aligned} \tag{7.3.7}$$

having noted that

$$\mathscr{U}_0(t, t_0)|n\rangle = e^{-i\omega_n(t-t_0)}|n\rangle. \tag{7.3.8}$$

The effect of $\mathscr{U}_0$ is merely a phase factor. Accordingly,

$$P_{m\to n} = |\langle n|\mathscr{U}_{\mathrm{I}}(t, t_0)|m\rangle|^2, \tag{7.3.9}$$

entirely in terms of the interaction factor in the evolution operator.

One can now make use of the series (7.2.15) for the evolution operator. The superscript on each term corresponds to the order of the perturbation calculation. Rapid convergence of the perturbation expansion will evidently be favoured, the better inequality (7.3.3) is fulfilled. To first order, the transition amplitude is given by

$$\langle n|\mathscr{U}_{\mathrm{I}}^{(1)}(t, t_0)|m\rangle = -\frac{i}{\hbar}\int_{t_0}^{t}\langle n|\mathscr{V}^{\mathrm{I}}(t')|m\rangle\,\mathrm{d}t'. \tag{7.3.10}$$

† P. A. M. Dirac, *Proc. Roy. Soc.* **A112**, 661 (1926); **A114**, 243 (1927).

Using (7.2.9) and the analogs of (7.3.8), we have

$$\langle n|\mathscr{V}^{\mathrm{I}}(t')|m\rangle = \langle n|\mathscr{U}_0^\dagger(t', t_0)\mathscr{V}(t')\mathscr{U}_0(t', t_0)|m\rangle$$
$$= e^{i\omega_{nm}(t'-t_0)} V_{nm}(t') \qquad (7.3.11)$$

with

$$\omega_{nm} \equiv \omega_n - \omega_m \qquad (7.3.12)$$

and

$$V_{nm}(t) \equiv \langle n|\mathscr{V}(t)|m\rangle. \qquad (7.3.13)$$

Therefore

$$\langle n|\mathscr{U}_{\mathrm{I}}^{(1)}(t, t_0)|m\rangle = -\frac{i}{\hbar} e^{-i\omega_{nm}t_0} \int_{t_0}^{t} e^{i\omega_{nm}t'} V_{nm}(t')\, dt'. \qquad (7.3.14)$$

The second-order contribution to transition amplitude is

$$\langle n|\mathscr{U}_{\mathrm{I}}^{(2)}(t, t_0)|m\rangle = -\hbar^2 \int_{t_0}^{t}\int_{t_0}^{t'} \langle n|\mathscr{V}^{\mathrm{I}}(t')\mathscr{V}^{\mathrm{I}}(t'')|m\rangle\, dt\, dt'. \qquad (7.3.15)$$

We can formally insert the resolution of the identity (cf. 3.9.17)

$$\mathbf{S}_k\, |k\rangle\langle k| = 1 \qquad (7.3.16)$$

between the operators $\mathscr{V}^{\mathrm{I}}(t')$ and $\mathscr{V}^{\mathrm{I}}(t'')$. Then, using (7.3.11),

$$\langle n|\mathscr{U}_{\mathrm{I}}^{(2)}(t, t_0)|m\rangle = -\hbar^{-2} e^{-i\omega_{nm}t_0}\, \mathbf{S}_k \int_{t_0}^{t}\int_{t_0}^{t'} e^{i\omega_{nk}t} e^{i\omega_{km}t''} V_{nk}(t') V_{km}(t'')\, dt'\, dt'. \qquad (7.3.17)$$

The transition probability (7.3.9) is given by

$$P_{m\to n} = |\delta_{nm} + \langle n|\mathscr{U}_{\mathrm{I}}^{(1)}|m\rangle + \langle n|\mathscr{U}_{\mathrm{I}}^{(2)}|m\rangle + \ldots|^2 \qquad (7.3.18)$$

To first order accuracy, for $n \neq m$,

$$P_{m\to n}^{(1)} = \left|\hbar^{-1}\int_{t_0}^{t} e^{i\omega_{nm}t'} V_{nm}(t')\, dt'\right|^2. \qquad (7.3.19)$$

If the first-order term vanishes (so that no cross-terms arise), the second-order transition probability is given by

$$P_{m\to n}^{(2)} = \left|\hbar^{-2}\, \mathbf{S}_k \int_{t_0}^{t}\int_{t_0}^{t'} e^{i\omega_{nk}t'} e^{i\omega_{km}t''} V_{nk}(t') V_{km}(t'')\, dt'\, dt''\right|^2. \qquad (7.3.20)$$

Higher-order contributions can be derived analogously.

When the perturbation matrix element $V_{nm} = 0$, the transition $m \to n$ is "forbidden" in first order. If, however, there exist states k such that both

V_{nk} and V_{km} are nonvanishing, the transition can occur in second order. Such transitions can be conceptualized as two-step processes through intermediate states k: $m \to k$, followed by $k \to n$. These intermediates in a higher-order transition are known as *virtual states*, since the usual laws restricting physical states (energy conservation, etc.) need not apply. Third-order transitions entail sequences like $m \to k \to l \to n$, through two virtual states. In general, nth order transitions involve $n - 1$ virtual states.

The structure of the transition probabilities (7.3.19), (7.3.20) and higher order analogs is such that, for nonzero $P^{(n)}$ and $P^{(m+m)}$,

$$\frac{P^{(n+m)}}{P^{(n)}} \sim \left(\frac{|V|\Delta t}{\hbar}\right)^{2m} \tag{7.3.21}$$

where V represents some typical perturbation matrix element and $\Delta t \equiv t - t_0$. The lowest nonvanishing order for a given transition is evidently a valid approximation whenever

$$\frac{|V_{\max}|\Delta t}{\hbar} \ll 1 \tag{7.3.22}$$

hence, for time intervals short according to

$$\Delta t \ll \hbar/|V_{\max}|. \tag{7.3.23}$$

Under these conditions, we also have that

$$P^{(n)} \ll 1, \tag{7.3.24}$$

showing that the initial state is only negligibly depleted during time Δt. As a matter of experience, contributions to a given transition above the lowest non-vanishing order do not generally improve the result, and may even give rise to convergence difficulties.

7.4 Alternative Derivation

The preceding perturbation formulas are derived in many textbooks by a somewhat less elegant approach, which does not make explicit use of interaction picture or the time-integral expansion. The alternative derivation is instructive nevertheless.

For the Hamiltonian (7.3.4), the time-dependent Schrödinger equation is given by

$$i\hbar\, \partial\Psi(q,t)/\partial t = [\mathscr{H}_0 + \mathscr{V}(t)]\,\Psi(q,t). \tag{7.4.1}$$

As discussed in Section 4.11 (cf. eqn 4.11.6), the states of this nonconservative system can be represented by expansions of the form

$$\Psi(q,t) = \mathop{\mathbf{S}}_{n} c_n(t)\,\psi_n(q) \tag{7.4.2}$$

in terms of the eigenstates of $\mathscr{H}_0$.

Substituting (7.4.2) into (7.4.1), with use of (7.3.2), we find

$$i\hbar \mathop{\mathsf{S}}_n \dot{c}_n(t)\psi_n(q) = \hbar \mathop{\mathsf{S}}_n c_n(t)\omega_n\psi_n(q) + \mathop{\mathsf{S}}_n c_n(t)\mathscr{V}(t)\psi_n(q). \tag{7.4.3}$$

Observe, however, that

$$i\hbar \frac{\mathrm{d}}{\mathrm{d}t}\left[c_n(t)\,\mathrm{e}^{i\omega_n t}\right] = i\hbar\dot{c}_n(t)\,\mathrm{e}^{i\omega_n t} - \hbar\omega_n c_n(t)\,\mathrm{e}^{i\omega_n t}. \tag{7.4.4}$$

Simplification is thereby effected by defining

$$C_n(t) \equiv c_n(t)\,\mathrm{e}^{i\omega_n t}, \tag{7.4.5}$$

which reduces (7.4.3) to

$$i\hbar \mathop{\mathsf{S}}_n \dot{C}_n(t)\,\mathrm{e}^{-i\omega_n t}\psi_n(q) = \mathop{\mathsf{S}}_n C_n(t)\,\mathrm{e}^{-i\omega_n t}\mathscr{V}(t)\psi_n(q). \tag{7.4.6}$$

The same result could have been obtained more directly by using the expansion (7.2.29), suggested by the interaction picture, in place of (7.4.2).

Taking scalar products successively with each $\psi_k(q)$ and using the notation (7.3.12) and (7.3.13), we obtain

$$i\hbar\dot{C}_k(t) = \mathop{\mathsf{S}}_n \mathrm{e}^{i\omega_{kn}t}V_{kn}(t)C_n(t). \tag{7.4.7}$$

This system of simultaneous linear differential equations for the time-dependent coefficients can also be expressed in matrix form as

$$i\hbar\,\mathrm{d}/\mathrm{d}t \begin{pmatrix} C_1(t) \\ C_2(t) \\ \vdots \end{pmatrix} = \begin{pmatrix} V_{11}(t) & V_{12}(t)\,\mathrm{e}^{i\omega_{12}t} & \cdots \\ V_{21}(t)\,\mathrm{e}^{i\omega_{21}t} & V_{22}(t) & \cdots \\ \cdots & \cdots & \cdots \end{pmatrix} \begin{pmatrix} C_1(t) \\ C_2(t) \\ \vdots \end{pmatrix} \tag{7.4.8}$$

This is, of course, the matrix equivalent of the interaction-picture Schrödinger equation (7.2.10). The above equations are, thus far, exact.

A common device in perturbation theory is to introduce a parameter λ $(0 \leqslant \lambda \leqslant 1)$ which can, in concept, turn the perturbation on or off in a continuous fashion. The Hamiltonian (7.3.4) is rewritten

$$\mathscr{H}(t) = \mathscr{H}_0 + \lambda\mathscr{V}(t) \tag{7.4.9}$$

such that $\lambda = 0$ and 1 correspond, respectively, to the unperturbed and perturbed systems. Equation (7.4.7) becomes accordingly

$$i\hbar\dot{C}_k(t) = \lambda \mathop{\mathsf{S}}_n \mathrm{e}^{i\omega_{kn}t}V_{kn}(t)C_n(t). \tag{7.4.10}$$

Each coefficient is now represented in a power series expansion:

$$C_k(t) = C_k^{(0)} + \lambda C_k^{(1)}(t) + \lambda^2 C_k^{(2)}(t) + \ldots \tag{7.4.11}$$

Assuming continuous behaviour of the system as λ varies between 0 and 1, the series (7.4.11) will behave as an analytic function of λ. One can therefore substitute (7.4.11) into (7.4.10) and equate coefficients of equal powers of λ. This results in the following set of equations:

$$\dot{C}_k^{(0)} = 0 \tag{7.4.12}$$

$$i\hbar\dot{C}_k^{(r)}(t) = \mathbf{S}_n\, e^{i\omega_{kn}t} V_{kn}(t) C_n^{(r-1)}(t), \qquad r = 1, 2, \ldots \tag{7.4.13}$$

The equations for each order r can, in principle, be integrated given the solutions of the next lower order, $r - 1$. Equations (7.4.12) show that the zeroth-order coefficients are constant in time, as is, indeed, to be expected, since the unperturbed problem is time-independent. The values of the $C_k^{(0)}$ serve as initial conditions in eqns (7.4.13). To relate the coefficients $C_n(t)$ to the transition amplitudes $\langle n|\mathscr{U}_I(t, t_0)|m\rangle$ (cf. 7.1.10) the initial state, at time t_0, is taken as the mth eigenstate of $\mathscr{H}_0$. This is effected by the conditions

$$\left.\begin{aligned} C_n^{(0)} &= \delta_{nm} \quad \text{or} \quad \delta(n - m) \\ C_n^{(r)}(t_0) &= 0, \quad n \neq m, \quad r = 1, 2, \ldots \end{aligned}\right\} \tag{7.4.14}$$

Using (7.4.14) in (7.4.13), with $r = 1$, and integrating, we obtain

$$C_k^{(1)}(t) = -\frac{i}{\hbar}\int_{t_0}^{t} e^{i\omega_{km}t'} V_{km}(t')\, dt'. \tag{7.4.15}$$

This corresponds to the first-order transition amplitude (7.3.14), apart from a trivial phase factor. For the second-order term ($r = 2$) change k to n in (7.4.15) and substitute into (7.4.13). After integrating, we have

$$C_k^{(2)}(t) = -\frac{1}{\hbar^2}\mathbf{S}_n \int_{t_0}^{t}\int_{t_0}^{t'} e^{i\omega_{kn}t'}\, e^{i\omega_{nm}t''} V_{kn}(t')\, V_{km}(t'')\, dt'\, dt'' \tag{7.4.16}$$

which corresponds to (7.3.17). Higher orders can be obtained analogously.

7.5 Constant Perturbation Turned on at t_0

A simple application of the preceding formulas is provided by the hypothetical case of a perturbation switched on sharply at time t_0 but independent of time thereafter, i.e.

$$\mathscr{V}(t) = \begin{cases} 0 & t < t_0 \\ \mathscr{V}_0 & t \geqslant t_0. \end{cases} \tag{7.5.1}$$

The Hamiltonian as a function of time is represented schematically in Fig. 7.1. The system, initially in the eigenstates Ψ_m of $\mathscr{H}_0$ evolves into a superposition of eigenstates $\mathbf{S}_n\, c_n(t)\Psi_n$ under the influence of the perturbation.

The perturbation thus induces transitions in the system in accordance with

$$P_{m\to n} = |c_n(t)|^2. \tag{7.5.2}$$

The results of this section apply equally well to a *transient perturbation*, in which $\mathscr{V}$ is subsequently *turned off* at the time $t_0 + \Delta t$. The transition probabilities at $t = t_0 + \Delta t$ would then pertain to *any* time $t \geqslant t_0 + \Delta t$.

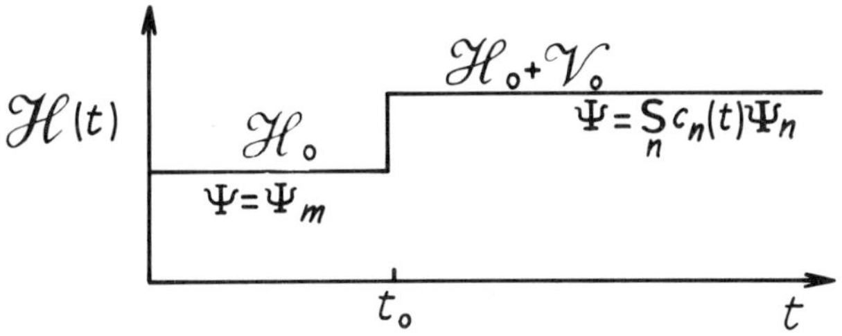

Fig. 7.1.

Using (7.5.1) in (7.3.19) we obtain for the first-order transition probability:

$$P^{(1)}_{m\to n} = \left| \hbar^{-1} V_{nm} \int_{t_0}^{t} e^{i\omega_{nm}t'}\, dt' \right|^2, \tag{7.5.3}$$

with the notation

$$V_{nm} \equiv \langle n|\mathscr{V}_0|m\rangle. \tag{7.5.4}$$

The integral can be evaluated explicitly since V_{nm} is time-independent:

$$\int_{t_0}^{t} e^{i\omega t'}\, dt' = \frac{e^{i\omega t} - e^{i\omega t_0}}{i\omega} = \frac{e^{i\omega(t+t_0)/2}}{i\omega}\left[e^{i\omega(t-t_0)/2} - e^{-i\omega(t-t_0)/2}\right]$$

$$= \frac{2e^{i\omega(t+t_0)/2}}{\omega} \sin \omega(t - t_0)/2. \tag{7.5.5}$$

Thus

$$P^{(1)}_{m\to n} = \frac{4|V_{nm}|^2}{\hbar^2\omega^2_{nm}} \sin^2 \tfrac{1}{2}\omega_{nm}\Delta t \tag{7.5.6}$$

where

$$\Delta t \equiv t - t_0. \tag{7.5.7}$$

For a purely discrete energy spectrum in the vicinity of E_m, the first-order transition probability exhibits the following rather quaint behaviour. For $E_n \neq E_m$, $P_{m\to n}$ oscillates as a function of t with period $\pi/|\omega_{nm}|$. For $E_n = E_m$, states degenerate with the initial state, $P_{m\to n}$ increases monotonically as $(\Delta t)^2$ (so long as $P_{m\to n} \ll 1$, so that first-order perturbation theory remains valid).

Of more physical significance is the situation in which the energy spectrum is continuous in the neighbourhood of E_m. The initial state m can itself belong to this continuum or it can be a discrete state immersed in the continuum. Let us, in any case, adopt the designation μ, ν for continuous quantum numbers.

It is convenient also to define the function

$$D(k, x) \equiv \frac{\sin^2 kx}{\pi k x^2} \tag{7.5.8}$$

wherein, with $k = \Delta t/2\hbar$, $x = E_\nu - E_\mu$, the transition probability density can be expressed

$$\rho^{(1)}_{\mu \to \nu} = \frac{2\pi \Delta t}{\hbar} |V_{\nu\mu}|^2 D\left(\frac{\Delta t}{2\hbar}, E_\nu - E_\mu\right). \tag{7.5.9}$$

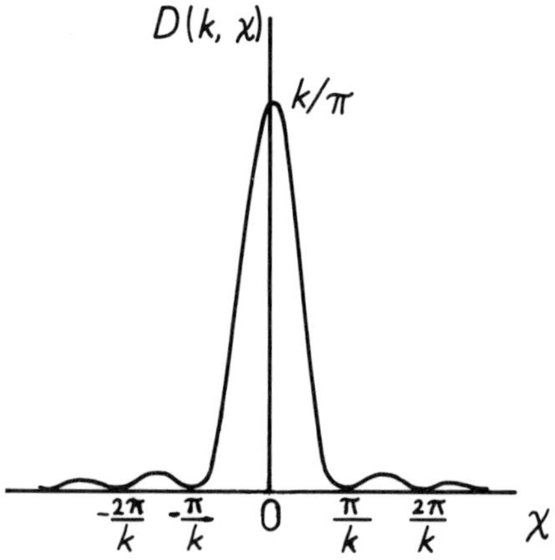

Fig. 7.2.

The D-function, plotted in Fig. 7.2, has the form of a finite slit diffraction pattern. In the limit $\Delta t \to \infty$ it becomes a deltafunction (cf. A.16):

$$\lim_{\Delta t \to \infty} D\left(\frac{\Delta t}{2\hbar}, E_\nu - E_\mu\right) = \delta(E_\nu - E_\mu). \tag{7.5.10}$$

While the perturbation can cause transitions to any state with $V_{\nu\mu} \neq 0$, the transition probability (7.5.9) is significantly different from zero only for final states ν falling within the central band of the D-function, within an energy $\pm h/\Delta t$ of the initial state μ. Thus

$$|E_\nu - E_\mu| \leqslant h/\Delta t, \tag{7.5.11}$$

which is approximately the energy bandwidth associated with the uncertainty principle (4.13.2). Since this energy band is centred about the initial state, energy is *approximately* conserved in the above transitions. The possible precision of energy specification is subject, however, to the limitations imposed by the uncertainty principle. Very fast processes—e.g., collisions of high-energy particles—can admit of quite a liberal leeway w.r.t. energy conservation, sufficient perhaps for the creation or annihilation of particles. The longer is Δt the more strict will be the energy-conservation condition.

From another point of view, perturbation such as (7.5.1) can be conceptualized as the interaction of the system with a measuring apparatus. In accord with the uncertainty principle, the longer the time interval Δt allotted for this measurement, the more accurately can the energy be known.

The side bands of the D-function are an artifice of the sudden turning-on of the perturbation at t_0. They have no essential physical significance and damp out rapidly as $\Delta t \to \infty$.

7.6 The Golden Rule

Many important physical phenomena are characterized by constant transition rates, meaning transition probabilities which increase linearly with time. One such instance is implicit in the continuum transitions considered above. The transition probability to each state in the energy-conserving band increases as $(\Delta t)^2$ while the bandwidth itself decreases as $1/\Delta t$. Thus the *composite* transition probability to the whole group of continuum states increases linearly with Δt.

To put this on a more quantitative basis, define

$$P_{\mu\to\{\nu\}} \equiv \int \rho_{\mu\to\nu}\, d\nu = \int \rho_{\mu\to\nu} g(E_\nu)\, dE_\nu, \tag{7.6.1}$$

in terms of the spectral density function $g(E_\nu)$ (cf. Section 5.3). From (7.5.9)

$$P^{(1)}_{\mu\to\{\nu\}} = \frac{2\pi\Delta t}{\hbar}\int |V_{\nu\mu}|^2 g(E_\nu) D\left(\frac{\Delta t}{2\hbar}, E_\nu - E_\mu\right) dE_\nu . \tag{7.6.2}$$

In the limit $\Delta t \to \infty$, the transitions become perfectly elastic (energy-conserving). By virtue of (7.5.10), we have, after integrating over the delta function,

$$P^{(1)}_{\mu\to\{\nu\}} = \frac{2\pi\Delta t}{\hbar}[|V_{\nu\mu}|^2 g(E_\nu)]_{E_\nu = E_\mu}. \tag{7.6.3}$$

Since this transition probability increases linearly with time, it is convenient to introduce the *transition rate*

$$W_{\mu\to\nu} \equiv \frac{dP_{\mu\to\nu}}{dt}. \tag{7.6.4}$$

Thus, to first order, the rate of elastic transitions from μ to the degenerate continuum states $\{\nu\}$ is given by

$$W^{(1)}_{\mu\to\{\nu\}} = \frac{2\pi}{\hbar}[|V_{\nu\mu}|^2 g(E_\nu)]_{E_\nu=E_\mu}. \tag{7.6.5}$$

This formula was designated the *golden rule* by Fermi†, in recognition of its extensive applicability. It is particularly useful for scattering problems, involving continua of unbound states. The following section shows its application to elastic scattering processes.

The golden rule can also be given in more abstract form by applying (7.5.10) directly to (7.5.9). The result is

$$W^{(1)}_{\mu\to\nu} = \frac{2\pi}{\hbar}|V_{\nu\mu}|^2\delta(E_\nu - E_\mu). \tag{7.6.6}$$

To get physically meaningful numbers, this must, of course, be integrated over a group of states.

One could alternatively have computed the transition rate directly from (7.5.6):

$$\begin{aligned} W^{(1)}_{m\to n} &= \frac{2|V_{nm}|^2}{\hbar^2\omega_{nm}}\sin\omega_{nm}\Delta t \\ &= \frac{2\pi}{\hbar}|V_{nm}|^2\frac{\sin(E_n - E_m)\Delta t/\hbar}{\pi(E_n - E_m)}. \end{aligned} \tag{7.6.7}$$

The last factor approaches a deltafunction as $\Delta t \to \infty$ (cf. A.12) and (7.6.7) reduces to (7.6.6) in the continuum case.

In the event that $V_{\nu\mu} = 0$ for all states ν in the immediate neighbourhood of μ, a second-order version of the golden rule is still applicable if there exist appropriate intermediate states κ such that $V_{\nu\kappa} \neq 0$ and $V_{\kappa\mu} \neq 0$. Putting (7.5.1) into the second-order formula (7.3.20) and carrying out the two integrations, there results

$$\rho^{(2)}_{\mu\to\nu} = \hbar^{-2}\left|\mathop{\mathsf{S}}_{\kappa}\frac{V_{\nu\kappa}V_{\kappa\mu}}{\omega_{\kappa\mu}}\left(\frac{e^{i\omega_{\nu\kappa}\Delta t} - 1}{\omega_{\nu\kappa}} - \frac{e^{i\omega_{\nu\mu}\Delta t} - 1}{\omega_{\nu\mu}}\right)\right|^2 \tag{7.6.8}$$

(having dropped the factor $e^{i\omega_{\nu\mu}t_0}$ of magnitude unity). For near elastic transitions, $\omega_{\nu\mu} \approx 0$ in the second denominator. By supposition (vanishing of first-order transition), no states with $\omega_{\kappa\mu} \approx 0$, hence with $\omega_{\nu\kappa} \approx 0$, will contribute to the summation. Thus the first term in (7.6.8) will never be "near resonance" and can be neglected in comparison with the second. Proceeding then in analogy with the derivation of (7.6.5), we obtain

$$W^{(2)}_{\mu\to\{\nu\}} = \frac{2\pi}{\hbar}\left[\left|\mathop{\mathsf{S}}_{\kappa}\frac{V_{\nu\kappa}V_{\kappa\mu}}{E_\kappa - E_\mu}\right|^2 g(E_\nu)\right]_{E_\nu=E_\mu}. \tag{7.6.9}$$

† E. Fermi *Nuclear Physics* (University of Chicago Press, 1950) p 142.

It is well to review the conditions of applicability of the golden rule. The time interval Δt during which the perturbation acts must be of such magnitude that

$$1/W_{\mu\rightarrow\{\nu\}} \gg \Delta t \gg \hbar/\Delta E \tag{7.6.10}$$

The first inequality guarantees that first (or lowest nonvanishing) perturbation order is adequate. In the second condition, ΔE is some representative energy measure for the system (e.g. e^2/a_0 for the hydrogen atom). The inequality then makes it valid to approximate the D-function by a delta-function, in accordance with (7.5.10) (or, physically, that energy conservation is adequately approximated). So long as (7.6.10) is fulfilled, the precise details as to how the perturbation is turned on and off become relatively insignificant. This flexibility extends the potential range of application of the golden-rule formalism.

7.7 Elastic Scattering Cross-section

The interaction of a free particle with a scattering centre can be viewed as a transient perturbation since a short-range potential $V(\mathbf{r})$ is effective only during the time the two particles are in close proximity. Although the cut-off times for the perturbation are not sharply defined, the golden rule can still be applied if the inequalities (7.6.10) are fulfilled.

Suppose a scattering event results in a transition from an initial plane-wave state $e^{i\mathbf{k}_0\cdot\mathbf{r}}$ to a final plane-wave state $e^{i\mathbf{k}\cdot\mathbf{r}}$, with $k = k_0$ for elastic scattering. It is appropriate to apply (7.6.5) with the spectral density function (cf. 5.8.5)

$$g(E, \theta, \phi) = \frac{mk}{(2\pi)^3\hbar^2} \tag{7.7.1}$$

representing the number of states per unit energy per unit solid angle. We have also set $V = 1$, signifying unit incident particle density.† Using (7.7.1) in (7.6.5), we obtain

$$j(\theta, \phi) = \frac{2\pi}{\hbar}\left|V_{\mathbf{k},\mathbf{k}_0}\right|^2 \frac{mk_0}{(2\pi)^3\hbar^2}\,. \tag{7.7.2}$$

Now, in the simplest instance, the golden rule gives the transition probability per unit time. Substitution of $g(E, \theta, \phi)$ in place of $g(E)$ converts it to scattering probability per unit time per unit solid angle. However, since free-particle eigenfunctions are fundamentally representations of many-particle beams (cf. Section 5.2), (7.7.2) can be identified with the *scattered particle flux*

† Alternatively, use of box-normalized eigenfunctions $V^{-\frac{1}{2}}e^{i\mathbf{k}\cdot\mathbf{r}}$ results in cancellation of the factor V in (5.8.5).

(cf. Section 5.9): the number of scattered particles per unit solid angle per unit time. Here (cf. 5.4.17)

$$V_{\boldsymbol{k},\boldsymbol{k}_0} = \int e^{-i\boldsymbol{k}.\boldsymbol{r}} V(\mathbf{r}) e^{i\boldsymbol{k}_0.\boldsymbol{r}} d^3\mathbf{r} = (2\pi)^3 V(\mathbf{K}) \quad (\mathbf{K} = \mathbf{k} - \mathbf{k}_0). \quad (7.7.3)$$

Dividing (7.7.2) by the incident flux

$$j_{\text{inc}} = \frac{\hbar k_0}{m} \quad (7.7.4)$$

we obtain the differential scattering cross-section

$$\sigma(\mathbf{K}) = \left| \frac{4\pi^2 m}{\hbar^2} V(\mathbf{K}) \right|^2. \quad (7.7.5)$$

We have thereby, from a completely different point of view, reproduced the result of the Born approximation (cf. 6.4.18).

7.8 Harmonic Perturbations

Perturbations containing oscillatory time dependence provide a prototype for the interaction of matter with radiation. Consider first a perturbation turned on at time t_0 of the following form

$$\mathscr{V}(t) = \begin{cases} 0, & t < t_0 \\ \mathscr{V}(\omega)\, e^{-i\omega t}, & t \geqslant t_0 \end{cases} \quad (7.8.1)$$

Such complex harmonic time dependence is representative of each monochromatic component of a Fourier-analyzed perturbation operator. Although (7.8.1) by itself is not hermitian, the following calculations correctly represent its contribution in a Fourier superposition.
Note that (7.5.1) is a particular case of (7.8.1), when $\omega = 0$.

Using (7.8.1) in (7.3.19), we obtain the first-order transition probability

$$P^{(1)}_{m\to n} = \left| \hbar^{-1} V_{nm}(\omega) \int_{t_0}^{t} e^{i(\omega_{nm} - \omega)t'}\, dt' \right|^2, \quad (7.8.2)$$

identical with (7.5.3) except for the exponential factor $(\omega_{nm} - \omega)$. Integrating and introducing the D-function, there results

$$P^{(1)}_{m\to n} = \frac{2\pi\Delta t}{\hbar} | V_{nm}(\omega)|^2 D\left(\frac{\Delta t}{2\hbar}, E_n - E_m - \hbar\omega\right) \quad (7.8.3)$$

$$= \frac{2\pi\Delta t}{\hbar^2} | V_{nm}(\omega)|^2 D\left(\frac{\Delta t}{2}, \omega_{nm} - \omega\right).$$

As we saw in Section 7.5, a constant perturbation $(\omega = 0)$ induced transitions which were preponderantly energy-conserving ($E_n \approx E_m$). In the present case, transitions occur to an energy band in which

$$E_n \approx E_m + \hbar\omega. \tag{7.8.4}$$

When $E_n > E_m$ ($\omega > 0$), this represents an *absorption* of energy by the system. If overall conservation of energy is assumed, this energy must come from the external system giving rise to the perturbation. If, on the other hand, $E_n < E_m$ ($\omega < 0$), then energy is being *lost* by the system. Such a process is known as *stimulated emission.* In either case, the transition energy is given by Bohr's frequency condition

$$\Delta E = \hbar\omega. \tag{7.8.5}$$

We have arrived, in fact, at a quantum-mechanical derivation of Bohr's famous postulate.

A physical realization of monochromaticity would, of course, require that $\mathscr{V}(t)$ be a hermitian operator. The perturbation (7.7.1) can be made hermitian by addition of its adjoint operator. We consider accordingly the modified time-dependent perturbation

$$\mathscr{V}(t) = \begin{cases} 0, & t < t_0 \\ \mathscr{V}(\omega)\,e^{-i\omega t} + \mathscr{V}^\dagger(\omega)\,e^{i\omega t}, & t \geqslant t_0. \end{cases} \tag{7.8.6}$$

Using (7.8.6) in the first-order perturbation formula, we calculate

$$\begin{aligned} P^{(1)}_{m\to n} &= \frac{1}{\hbar^2}\left| V_{nm}(\omega)\int_{t_0}^{t} e^{i(\omega)_{nm}-\omega)t'}\,dt' + V^*_{mn}(\omega)\int_{t_0}^{t} e^{i(\omega_{nm}+\omega)t'}\,dt'\right|^2 \\ &= \frac{4}{\hbar^2}\left| V_{nm}(\omega)\,e^{i(\omega_{nm}-\omega)(t+t_0)/2}\,\frac{\sin\frac{1}{2}(\omega_{nm}-\omega)\Delta t}{\omega_{nm}-\omega}\right. \\ &\qquad \left. + V^*_{mn}(\omega)\,e^{i(\omega_{nm}+\omega)(t+t_0)/2}\,\frac{\sin\frac{1}{2}(\omega_{nm}+\omega)\Delta t}{\omega_{nm}+\omega}\right|^2 \end{aligned} \tag{7.8.7}$$

having used analogs of (7.5.5). This expression will have significant magnitude only if one of the resonance denominators is near zero. When $\omega \approx \omega_{nm}$, the first term within the absolute value bars will dominate and the second can be neglected. In terms of the D-function, we then have

$$P^{(1)}_{m\to n} = \frac{2\pi\Delta t}{\hbar^2}|V_{nm}(\omega)|^2 D\left(\frac{\Delta t}{2}, \omega_{nm} - \omega\right). \tag{7.8.8}$$

If, on the other hand, $\omega \approx -\omega_{nm}$, the second term in (7.8.9) dominates and

$$P^{(1)}_{m\to n} = \frac{2\pi\Delta t}{\hbar^2}|V_{mn}(\omega)|^2 D\left(\frac{\Delta t}{2}, \omega_{nm} + \omega\right). \tag{7.8.9}$$

From a physical point of view, the first-order transition probability is non-negligible only if the perturbation frequency ω is close to some resonance frequency of the system in accordance with the Bohr condition (7.8.5). For $E_n > E_m$, (7.8.8) corresponds to absorption. Interchanging quantum numbers m and n in (7.8.9) and noting that $D(k, x)$ is an even function of x, we find

$$P^{(1)}_{n\to m} = \frac{2\pi\Delta t}{\hbar^2}|V_{nm}(\omega)|^2 D\left(\frac{\Delta t}{2}, \omega_{nm} - \omega\right). \tag{7.8.10}$$

This represents the probability of stimulated emission from n to m. Comparing (7.8.10) with (7.8.8), the probabilities of absorption and stimulated emission between two given levels are seen to be exactly equal. This has to be shown specifically for a first-order allowed transition ($V_{nm}(\omega) \neq 0$) but is, in fact, an instance of the more general principle (7.1.16).

To generalize further, we consider a polychromatic perturbation made up of a continuous superposition of harmonic components (7.8.1). This can be represented by

$$\mathscr{V}(t) = \int_{-\infty}^{\infty} \mathscr{V}(\omega)\, e^{-i\omega t}\, d\omega \tag{7.8.11}$$

in which

$$\mathscr{V}(-\omega) = \mathscr{V}^{\dagger}(\omega) \tag{7.8.12}$$

in order that $\mathscr{V}(t)$ be hermitian. For concreteness, we assume a transient perturbation during the finite time interval Δt from t_0 to t; thus the Fourier components $\mathscr{V}(\omega)$ are to be consistent with

$$\mathscr{V}(t') = 0 \quad \text{for} \quad t' < t_0 \quad \text{and} \quad t' > t\,. \tag{7.8.13}$$

By (7.3.19)

$$P^{(1)}_{m\to n} = \hbar^{-2}\left|\int_{t_0}^{t} V_{nm}(t')\, e^{i\omega_{nm}t'}\, dt'\right|^2. \tag{7.8.14}$$

But, by virtue of (7.8.13), the limits of integration can be extended to $\pm\infty$. The integral then becomes the Fourier transform of $V_{nm}(t)$ (cf. B.24):

$$\int_{-\infty}^{\infty} V_{nm}(t')\, e^{i\omega_{nm}t'}\, dt' = 2\pi V_{nm}(\omega_{nm}). \tag{7.8.15}$$

We have written

$$V_{nm}(\omega_{nm}) \equiv \langle n|\mathscr{V}(\omega_{nm})|m\rangle \tag{7.8.16}$$

for the Fourier component of the perturbation matrix element at the resonance frequency $\omega = \omega_{nm}$.

Thus the transition probability for (7.8.11) acting over the time interval

Δt is

$$P^{(1)}_{m\to n} = \frac{4\pi^2}{\hbar^2}\,|V_{nm}(\omega_{nm})|^2 \tag{7.8.17}$$

and the average transition probability per unit time is given by†

$$W^{(1)}_{m\to n} = \frac{4\pi^2}{\hbar^2\Delta t}\,|V_{nm}(\omega_{nm})|^2. \tag{7.8.18}$$

These formulas apply both to absorption, $\omega_{nm} > 0$, and to stimulated emission, $\omega_{nm} < 0$. In the following chapter we shall specialize this result to treat the absorption and emission of radiation by atomic systems.

† This result, in fact, subsumes all our previous work on transient perturbations. Thus, for a constant perturbation (7.5.1) and (7.8.18) reduces to (7.5.6)

$$|V_{nm}(\omega)|^2 = \frac{1}{\pi^2\omega^2}\,|V_{nm}|^2 \sin^2 \tfrac{1}{2}\omega\Delta t.$$

8

Matter and Radiation

Phenomena in chemistry and atomic physics involving atoms or molecules in electromagnetic fields can be quite satisfactorily accounted for by a semiclassical theory of radiation. The theory is *semiclassical* in the sense that atomic particles governed by the laws of quantum theory are assumed to interact with electromagnetic fields conforming to the classical Maxwell–Lorentz equations. A more correct and consistent theory would, of course, take into account the quantum nature of electromagnetic fields as assemblies of photons. Yet, in the low-energy domain, most of the results of semiclassical radiation theory agree with those of quantum electrodynamics. One important defect is spontaneous emission, which is predicted only by the quantum theory of radiation. Still, an appropriate *ad hoc* adaptation from classical radiation theory accounts for even this phenomenon within a semiclassical context.

8.1 Schrödinger Equation in an Electromagnetic Field

In classical nonrelativistic theory, a charged particle in an electromagnetic field is represented by the Hamiltonian function (cf. 2.6.19)

$$H = \frac{\pi^2}{2m} + q\Phi \tag{8.1.1}$$

in which

$$\boldsymbol{\pi} \equiv \mathbf{p} - \frac{q}{c}\mathbf{A} \tag{8.1.2}$$

is the kinematical momentum

$$\boldsymbol{\pi} = m\mathbf{v}. \tag{8.1.3}$$

The corresponding quantum-mechanical Hamiltonian operator has the form

$$\mathscr{H} = \tfrac{1}{2}m\Pi^2 + q\Phi \tag{8.1.4}$$

in terms of the *kinematical momentum operator*

$$\mathbf{\Pi} \equiv -i\hbar\nabla - \frac{q}{c}\mathbf{A}. \tag{8.1.5}$$

The square of this operator has the explicit form

$$\Pi^2 = -\hbar^2\nabla^2 + \frac{i\hbar q}{c}(\nabla\cdot\mathbf{A} + \mathbf{A}\cdot\nabla) + \frac{q^2}{c^2}A^2. \tag{8.1.6}$$

Now

$$\nabla\cdot\mathbf{A} = (\nabla\cdot\mathbf{A}) + \mathbf{A}\cdot\nabla \tag{8.1.7}$$

in which operation on some wavefunction is implicit. The notation $(\nabla\,.\,\mathbf{A})$ signifies that the operator ∇ in this term does *not* get through to the wavefunction†. We find accordingly

$$\mathscr{H} = -\frac{\hbar^2}{2m}\nabla^2 + \frac{i\hbar q}{mc}\mathbf{A}\cdot\nabla + \frac{i\hbar q}{2mc}(\nabla\cdot\mathbf{A}) + \frac{q^2}{2mc^2}A^2 + q\Phi. \tag{8.1.8}$$

When q equals the electronic charge $-e$, one can introduce the fine-structure constant

$$\alpha \equiv e^2/\hbar c \approx 1/137. \tag{8.1.9}$$

The Hamiltonian can thereby be written

$$\mathscr{H} = \mathscr{H}^{(0)} + \alpha\mathscr{H}^{(1)} + \alpha^2\mathscr{H}^{(2)} \tag{8.1.10}$$

with

$$\mathscr{H}^{(0)} = -\frac{\hbar^2}{2m}\nabla^2 - e\Phi \tag{8.1.11}$$

$$\mathscr{H}^{(1)} = -\frac{i\hbar^2}{me}[\mathbf{A}\cdot\nabla + \tfrac{1}{2}(\nabla\cdot\mathbf{A})] \tag{8.1.12}$$

$$\mathscr{H}^{(2)} = \frac{\hbar^2}{2me^2}A^2. \tag{8.1.13}$$

The magnitudes of successive orders are in the approximate ratio

$$\left|\frac{\langle\mathscr{H}^{(1)}\rangle}{\langle\mathscr{H}^{(0)}\rangle}\right| \sim \left|\frac{\langle\mathscr{H}^{(2)}\rangle}{\langle\mathscr{H}^{(1)}\rangle}\right| \sim \frac{eA}{cp} \tag{8.1.14}$$

where $\mathbf{p}$ is the particle momentum.

A charged particle in an electromagnetic field is governed by the time-

† This can also be expressed in commutator form:

$$(\nabla\cdot\mathbf{A}) = \nabla\cdot\mathbf{A} - \mathbf{A}\cdot\nabla = \sum_i [\nabla_i, A_i].$$

dependent Schrödinger equation

$$i\hbar\frac{\partial}{\partial t}\Psi(\mathbf{r}, t) = \mathscr{H}\Psi(\mathbf{r}, t) \tag{8.1.15}$$

with the Hamiltonian (8.1.8). It is interesting to note that, if the *retarded* forms of the electromagnetic potentials (cf. eqns 6.1.57, 59) are used, the Hamiltonian (8.1.8) does *not* fulfil the condition (4.7.3), i.e.,

$$\mathscr{H}(-t) \neq \mathscr{H}^*(t). \tag{8.1.16}$$

Thereby the time-inversion symmetry (4.7.5) does not pertain. Suppose, however, that one takes the following linear combinations of the retarded and advanced potentials:

$$\begin{aligned} \mathbf{A}(\mathbf{r}, t) &= \tfrac{1}{2}[\mathbf{A}_{\text{ret}}(\mathbf{r}, t) - \mathbf{A}_{\text{adv}}(\mathbf{r}, t)] \\ \Phi(\mathbf{r}, t) &= \tfrac{1}{2}[\Phi_{\text{ret}}(\mathbf{r}, t) + \Phi_{\text{adv}}(\mathbf{r}, t)] \end{aligned} \tag{8.1.17}$$

Then†

$$\mathbf{A}(\mathbf{r}, -t) = -\mathbf{A}(\mathbf{r}, t), \qquad \Phi(\mathbf{r}, -t) = \Phi(\mathbf{r}, t) \tag{8.1.18}$$

such as to make (8.1.8) conform to

$$\mathscr{H}(-t) = \mathscr{H}^*(t). \tag{8.1.19}$$

Thereby the Schrödinger equation (8.1.15) for a particle in an electromagnetic field becomes symmetrical with respect to time inversion.

8.2 Gauge Invariance

It will next be demonstrated that the time-dependent Schrödinger equation (8.1.15) is invariant under gauge transformation. The alternative potentials (cf. 2.2.7)

$$\mathbf{A}' = \mathbf{A} + \nabla\chi, \qquad \Phi' = \Phi - \frac{1}{c}\frac{\partial\chi}{\partial t} \tag{8.2.1}$$

transform the Schrödinger equation to

$$i\hbar\frac{\partial\Psi'}{\partial t} = \mathscr{H}'\Psi' \tag{8.2.2}$$

with

$$\mathscr{H}' = \frac{1}{2m}\Pi'^2 + e\Phi' \tag{8.2.3}$$

† In 4-vector notation, (8.1.18) can be expressed

$$A_\nu^*(x) = -A_\nu(x) \quad \nu = 1\ldots4$$

which is analogous to the conjugation relation for the momentum operator: $p_\nu^* = -p_\nu$.

$$\mathbf{\Pi}' = -i\hbar\nabla - \frac{q}{c}\mathbf{A}' = \mathbf{\Pi} - \frac{q}{c}\nabla\chi. \tag{8.2.4}$$

To find the form of the gauge-transformed wavefunction Ψ', we note first the operator relation

$$-i\hbar\nabla\, e^{iq\chi/\hbar c} = e^{iq\chi/\hbar c}\left[\frac{q}{c}\nabla\chi - i\hbar\nabla\right]. \tag{8.2.5}$$

By inserting $(q/c)\mathbf{A}$ on each side of (8.2.5) and using (8.2.4), we arrive at the identity

$$\mathbf{\Pi}'\, e^{iq\chi/\hbar c} = e^{iq\chi/\hbar c}\,\mathbf{\Pi}. \tag{8.2.6}$$

Applying $\mathbf{\Pi}'$ to (8.2.6):

$$\Pi'^2\, e^{iq\chi/\hbar c} = \mathbf{\Pi}'\cdot e^{iq\chi/\hbar c}\,\mathbf{\Pi} = \mathbf{\Pi}'\, e^{iq\chi/\hbar c}\cdot\mathbf{\Pi} = e^{iq\chi/\hbar c}\,\Pi^2. \tag{8.2.7}$$

Analogously, we find

$$i\hbar\frac{\partial}{\partial t}\, e^{iq\chi/\hbar c} = e^{iq\chi/\hbar c}\left[-\frac{q}{c}\left(\frac{\partial\chi}{\partial t}\right) + i\hbar\frac{\partial}{\partial t}\right] \tag{8.2.8}$$

and, by inserting $q\Phi$ in both sides,

$$\left(i\hbar\frac{\partial}{\partial t} - q\Phi'\right) e^{iq\chi/\hbar c} = e^{iq\chi/\hbar c}\left(i\hbar\frac{\partial}{\partial t} - q\Phi\right). \tag{8.2.9}$$

Combining (8.2.9) with (8.2.7) such as to form the operators (8.1.4) and (8.2.3), we obtain

$$\left(i\hbar\frac{\partial}{\partial t} - \mathscr{H}'\right) e^{iq\chi/\hbar c} = e^{iq\chi/\hbar c}\left(i\hbar\frac{\partial}{\partial t} - \mathscr{H}\right). \tag{8.2.10}$$

The right-hand side applied to some solution $\Psi(\mathbf{r}, t)$ of the original time-dependent Schrödinger equation (8.1.15) gives zero. Thus

$$\left(i\hbar\frac{\partial}{\partial t} - \mathscr{H}'\right) e^{iq\chi/\hbar c}\,\Psi = 0 \tag{8.2.11}$$

so that the gauge-transformed wavefunction must be given by

$$\Psi'(\mathbf{r}, t) = e^{iq\chi(\mathbf{r}, t)/\hbar c}\,\Psi(\mathbf{r}, t). \tag{8.2.12}$$

Since Ψ' differs from Ψ only by a phase factor, the two are physically equivalent. In particular,

$$|\Psi'(\mathbf{r}, t)|^2 = |\Psi(\mathbf{r}, t)|^2. \tag{8.2.13}$$

It has thereby been demonstrated that the time-dependent Schrödinger equation (8.1.15) is invariant under the extended gauge transformation augmenting (8.2.1) by (8.2.12).

The time-independent Schrödinger equation

$$\left(\frac{1}{2m}\Pi^2 + q\Phi\right)\psi(\mathbf{r}) = E\psi(\mathbf{r}) \tag{8.2.14}$$

can apply in the case of a *static* field: **A** and Φ independent of time. Any gauge function $\chi(\mathbf{r})$ should evidently be time-independent as well. Under these circumstances, the Schrödinger equation gauge transforms to

$$\mathscr{H}'\psi'(\mathbf{r}) = E'\psi'(\mathbf{r}) \tag{8.2.15}$$

such that

$$\begin{aligned} \mathbf{A}'(\mathbf{r}) &= \mathbf{A}(\mathbf{r}) + \nabla\cdot\chi(\mathbf{r}) \\ \Phi'(\mathbf{r}) &= \Phi(\mathbf{r}) \\ \psi'(\mathbf{r}) &= e^{iq\chi(\mathbf{r})/\hbar c}\,\psi(\mathbf{r}) \end{aligned} \tag{8.2.16}$$

and

$$E' = E.$$

8.3 Continuity Equation

The continuity equation obtained in Section 4.4 must be generalized in the presence of a vector potential. By an analogous derivation, one premultiplies (8.1.15) by Ψ^* and subtracts the complex conjugate equation. This gives

$$i\hbar\left(\Psi^*\frac{\partial\Psi}{\partial t} + \Psi\frac{\partial\Psi^*}{\partial t}\right) = \Psi^*\mathscr{H}\Psi - \Psi(\mathscr{H}\Psi)^*. \tag{8.3.1}$$

Making use of (8.1.8), the right-hand side works out to

$$\begin{aligned} &-\frac{\hbar^2}{2m}(\Psi^*\nabla^2\Psi - \Psi\nabla^2\Psi^*) + \frac{i\hbar q}{mc}[\Psi^*\mathbf{A}\cdot\nabla\Psi + \Psi\mathbf{A}\cdot\nabla\Psi^* + (\nabla\cdot\mathbf{A})\Psi^*\Psi] \\ &\qquad = \nabla\cdot\left[-\frac{\hbar^2}{2m}(\Psi^*\nabla\Psi - \Psi\nabla\Psi^*) + \frac{i\hbar q}{mc}\Psi^*\mathbf{A}\Psi\right] \end{aligned} \tag{8.3.2}$$

Thus, after division by $i\hbar$, (8.3.1) transforms to an equation of continuity

$$\frac{\partial\rho}{\partial t} + \nabla\cdot\mathbf{j} = 0. \tag{8.3.3}$$

The probability density is

$$\rho(\mathbf{r}, t) = \Psi^*(\mathbf{r}, t)\,\Psi(\mathbf{r}, t) \tag{8.3.4}$$

just as in the field-free case (cf. 4.4.4). For the current density, however,

(cf. 4.4.5)

$$\mathbf{j}(\mathbf{r}, t) = -\frac{i\hbar}{2m}(\Psi^*\nabla\Psi - \Psi\nabla\Psi^*) - \frac{q}{mc}\Psi^*\mathbf{A}\Psi. \tag{8.3.5}$$

which contains an additional term contributed by the vector potential. Alternatively, by introducing the kinematical momentum operator (8.1.5), we can write

$$\mathbf{j}(\mathbf{r}, t) = \frac{1}{2m}[\Psi^*\mathbf{\Pi}\Psi + \Psi(\mathbf{\Pi}\Psi)^*]. \tag{8.3.6}$$

8.4 Ehrenfest Relations

It is also of interest to work out the Ehrenfest relations (Section 4.6) for the specific case of a particle in an electromagnetic field. The Heisenberg equation for the x-coordinate reads

$$\frac{\mathrm{d}\langle x\rangle}{\mathrm{d}t} = (i\hbar)^{-1}\langle[x, \mathscr{H}]\rangle. \tag{8.4.1}$$

For the Hamiltonian (8.1.4)

$$[x, \mathscr{H}] = \frac{1}{2m}[x, \Pi_x^2] = \frac{1}{2m}(\Pi_x[x, \Pi_x] + [x, \Pi_x]\Pi_x). \tag{8.4.2}$$

But

$$[x, \Pi_x] = [x, p_x] = i\hbar \tag{8.4.3}$$

which leads to

$$\frac{\mathrm{d}\langle x\rangle}{\mathrm{d}t} = \frac{1}{m}\langle\Pi_x\rangle. \tag{8.4.4}$$

The vector generalization is

$$\frac{\mathrm{d}\langle \mathbf{r}\rangle}{\mathrm{d}t} = \frac{1}{m}\langle\mathbf{\Pi}\rangle \tag{8.4.5}$$

and this, of course, corresponds to the classical relation

$$\boldsymbol{\pi} = m\mathbf{v}. \tag{8.4.6}$$

The kinematical momentum obeys the equation of motion

$$\frac{\mathrm{d}\langle\Pi_x\rangle}{\mathrm{d}t} = \left\langle\frac{\partial\Pi_x}{\partial t}\right\rangle + (i\hbar)^{-1}\langle[\Pi_x, \mathscr{H}]\rangle. \tag{8.4.7}$$

The first term on the right comes from the time dependence of the vector potential: in fact

$$\frac{\partial \Pi_x}{\partial t} = -\frac{q}{c}\frac{\partial A_x}{\partial t}. \tag{8.4.8}$$

For the commutator,

$$[\Pi_x, \mathscr{H}] = \frac{1}{2m}[\Pi_x, \Pi^2] + q[\Pi_x, \Phi]. \tag{8.4.9}$$

Now

$$[\Pi_x, \Phi] = -i\hbar\frac{\partial \Phi}{\partial x} \tag{8.4.10}$$

while

$$[\Pi_x, \Pi^2] = [\Pi_x, \Pi_y]\Pi_y + \Pi_y[\Pi_x, \Pi_y] + [\Pi_x, \Pi_z]\Pi_z + \Pi_z[\Pi_x, \Pi_z]. \tag{8.4.11}$$

We can apply the following commutation relations:†

$$[\Pi_x, \Pi_y] = \frac{i\hbar q}{c}H_z \quad \text{et cyc} \tag{8.4.12}$$

where **H** is the magnetic field

$$\mathbf{H} = \nabla \times \mathbf{A}. \tag{8.4.13}$$

Recalling also that

$$\mathbf{E} = -\frac{1}{c}\frac{\partial \mathbf{A}}{\partial t} - \nabla\Phi \tag{8.4.14}$$

we obtain for (8.4.7)

$$\frac{\mathrm{d}\langle \Pi_x \rangle}{\mathrm{d}t} = q\langle E_x \rangle + \frac{q}{2mc}\langle H_z\Pi_y + \Pi_y H_z - H_y\Pi_z - \Pi_z H_y \rangle. \tag{8.4.15}$$

Thus

$$\frac{\mathrm{d}\langle \boldsymbol{\Pi} \rangle}{\mathrm{d}t} = q\langle \mathbf{E} \rangle + \frac{q}{2mc}\langle \boldsymbol{\Pi} \times \mathbf{H} - \mathbf{H} \times \boldsymbol{\Pi} \rangle. \tag{8.4.16}$$

This corresponds, by virtue of (8.4.6), to the classical nonrelativistic Lorentz

† These can also be expressed in symbolic vector form as

$$\boldsymbol{\Pi} \times \boldsymbol{\Pi} = \frac{i\hbar q}{c}\mathbf{H}.$$

force equation (cf. 2.5.17)

$$m\frac{d\mathbf{v}}{dt} = q\mathbf{E} + \frac{q}{c}\mathbf{v} \times \mathbf{H}. \tag{8.4.17}$$

8.5 Radiative Transitions

Among the most important processes in nature are transitions among stationary states of atoms and molecules induced by electromagnetic radiation. We can now calculate the appropriate transition probabilities, having developed the general theory of transitions and the specific forms of the electromagnetic interaction operators. Consider for simplicity an unperturbed one-electron system (we will generalize later) described by a time-independent Hamiltonian

$$\mathscr{H}^{(0)} = -\frac{\hbar^2}{2m}\nabla^2 + V(\mathbf{r}). \tag{8.5.1}$$

We presume to know all the stationary states of $\mathscr{H}^{(0)}$, namely the solutions of

$$\mathscr{H}^{(0)}\psi_n(\mathbf{r}) = E_n\psi_n(\mathbf{r}) \tag{8.5.2}$$

for all eigenvalues and eigenfunctions. Let an electromagnetic field $\mathbf{A}$, $\mathbf{\Phi}$ now be applied to the system. The Hamiltonian (8.1.8) now pertains, with the additional term $V(\mathbf{r})$ and with $q = -e$:

$$\mathscr{H} = \mathscr{H}^{(0)} - \frac{ie\hbar}{mc}\mathbf{A}\cdot\mathbf{\nabla} - \frac{ie\hbar}{2mc}(\mathbf{\nabla}\cdot\mathbf{A}) + \frac{e^2}{2mc^2}A^2 - e\Phi\,. \tag{8.5.3}$$

Assuming more specifically, that the radiation is in the form of transverse plane waves, we have by (2.10.2)

$$\Phi = 0, \qquad (\mathbf{\nabla}\cdot\mathbf{A}) = 0 \tag{8.5.4}$$

thus eliminating the third and fifth terms of (8.5.3). To compare the magnitudes of the two remaining perturbation terms, note that a typical spectroscopic source intensity might be of the order of $I \sim 100$ watts/cm^2 $= 10^9$ ergs/cm^2 sec. The intensity is related to the vector potential by (2.10.15):

$$I = \omega^2\bar{A}^2/4\pi c. \tag{8.5.5}$$

Thus, for visible light with $\omega \sim 5 \times 10^{15}$ sec^{-1}, $A \sim 5 \times 10^{-6}$ esu/cm. For an atomic scale system, the operator $\mathbf{\nabla}$ has an expectation value of the order of $1/a_0 \sim 10^8$ cm^{-1} ($a_0 \equiv \hbar^2/me^2 = 0{\cdot}529 \times 10^{-8}$ cm, the Bohr radius). We find thereby

$$\left|\frac{\left\langle \frac{e^2}{2mc^2}A^2\right\rangle}{\left\langle -\frac{ie\hbar}{mc}\mathbf{A}\cdot\mathbf{\nabla}\right\rangle}\right| \sim 10^{-7} \tag{8.5.6}$$

Moreover, since $|\langle \mathscr{H}^{(0)} \rangle| \sim \hbar^2/2ma_0^2$ for an atomic system,

$$\left| \frac{\left\langle -\frac{ie\hbar}{mc} \mathbf{A} \cdot \nabla \right\rangle}{\langle \mathscr{H}^{(0)} \rangle} \right| \sim 10^{-7} \tag{8.5.7}$$

These magnitudes show that, under ordinary circumstances, the A^2 term is negligible compared to the $\mathbf{A} \cdot \nabla$ term and that the latter can be validly treated as a perturbation on $\mathscr{H}^{(0)}$. We can accordingly simplify (8.5.3) to

$$\mathscr{H} = \mathscr{H}^{(0)} + \mathscr{V} \tag{8.5.8}$$

with the perturbation operator

$$\mathscr{V} = -\frac{ie\hbar}{mc} \mathbf{A} \cdot \nabla \tag{8.5.9}$$

representing the essential portion of the interaction of the system with a radiation field.

The results of Section 7.8 can be applied directly if the radiation consists of a superposition of plane waves, as represented by the vector potential (cf. 2.10.16):

$$\mathbf{A}(\mathbf{r}, t) = \int_{-\infty}^{\infty} \mathscr{A}(\omega) \, \mathrm{e}^{i(\mathbf{k} \cdot \mathbf{r} - \omega t)} \, \mathrm{d}\omega. \tag{8.5.10}$$

For the perturbation operator expressed in the form (cf. 7.8.11)

$$\mathscr{V}(t) = \int_{-\infty}^{\infty} \mathscr{V}(\omega) \, \mathrm{e}^{-i\omega t} \, \mathrm{d}\omega \tag{8.5.11}$$

the Fourier component of frequency ω is given by

$$\mathscr{V}(\omega) = -\frac{ie\hbar}{mc} \mathrm{e}^{i\mathbf{k} \cdot \mathbf{r}} \mathscr{A}(\omega) \cdot \nabla. \tag{8.5.12}$$

A matrix element at a resonance frequency

$$\omega = \omega_{nm} = (E_n - E_m)/\hbar \tag{8.5.13}$$

$$k = |\mathbf{k}| = \omega_{nm}/c \tag{8.5.14}$$

is accordingly

$$V_{nm}(\omega_{nm}) = -\frac{ie\hbar}{mc} \mathscr{A}(\omega_{nm}) \cdot \langle n | \mathrm{e}^{i\mathbf{k} \cdot \mathbf{r}} \, \nabla | m \rangle. \tag{8.5.15}$$

This can be written

$$V_{nm}(\omega_{nm}) = -\frac{ie\hbar}{mc}\mathscr{A}(\omega_{nm})\langle n|e^{i\mathbf{k}\cdot\mathbf{r}}\,\mathbf{e}\cdot\nabla|m\rangle \tag{8.5.16}$$

where $\mathbf{e}$ is the unit vector in the direction of polarization.

The first-order transition rate for absorption or stimulated emission of radiation can now be given. By (7.8.18)

$$W_{mn} = \frac{4\pi^2}{\hbar^2\Delta t}|V_{nm}(\omega_{nm})|^2. \tag{8.5.17}$$

Using (8.5.16) and the radiation intensity relation (2.10.25)

$$I(\omega) = \omega^2|\mathscr{A}(\omega)|^2/c\Delta t \tag{8.5.18}$$

we obtain

$$W_{mn} = \frac{4\pi^2 e^2}{m^2 c\omega_{nm}^2} I(\omega_{nm})|\langle n|e^{i\mathbf{k}\cdot\mathbf{r}}\,\mathbf{e}\cdot\nabla|m\rangle|^2. \tag{8.5.19}$$

This applies both to absorption, $\omega_{nm} > 0$, and to stimulated emission, $\omega_{nm} < 0$. In absorption, a photon of energy $\hbar\omega_{nm}$ is taken up by the system; in stimulated emission a photon $\hbar\omega_{nm}$ knocks out an identical photon. In both cases, the transition rate is proportional to the intensity of incident radiation.

To reiterate, eqn (8.5.18) is valid provided that the radiation is not excessively intense and that first-order perturbation theory is adequate. This limits our consideration, according to quantum electrodynamics, to single-photon processes. For intense laser sources, the A^2 term, as well as second-order contributions from the $\mathbf{A}\cdot\nabla$ term, may become significant, thus giving rise to two-photon processes,

8.6 Electric Dipole Transitions

In radiative processes involving atoms and molecules, the radiation wavelength is typically many times greater than the linear dimensions of the system:

$$\lambda \gg a. \tag{8.6.1}$$

For example $a \sim 10^{-8}$ cm for an atom while $\lambda \sim 10^{-5}$ cm (1000 A) for ultraviolet radiation. The magnitude of the wavevector is

$$k = 2\pi/\lambda \tag{8.6.2}$$

so that

$$|\mathbf{k}\cdot\mathbf{r}| \sim 10^{-2}. \tag{8.6.3}$$

It is thus reasonable to expand the exponential factor $e^{i\mathbf{k}\cdot\mathbf{r}}$:

$$e^{i\mathbf{k}\cdot\mathbf{r}} = 1 + i\mathbf{k}\cdot\mathbf{r} - \tfrac{1}{2}(\mathbf{k}\cdot\mathbf{r})^2 - \ldots \tag{8.6.4}$$

and to keep only the leading term for which the matrix element does not vanish.† The lowest order approximation consists in simply replacing the exponential by unity:

$$W_{mn}^{\text{dipole}} = \frac{4\pi^2 e^2}{m^2 c\omega_{nm}^2} I(\omega_{nm})|\langle n|\mathbf{e}\cdot\boldsymbol{\nabla}|m\rangle|^2. \tag{8.6.5}$$

The preceding development is directly applicable to the interaction of polarized radiation with an oriented single crystal. When the sample is a gas, a liquid or a crystalline powder, molecular orientations are randomized with respect to the direction of polarization. The appropriate transition rate in these cases is the ensemble average of (8.6.5) over all molecular orientations. Now, with respect to some instantaneous molecule-fixed coordinate system,

$$\langle n|\mathbf{e}\cdot\boldsymbol{\nabla}|m\rangle = \sum_{i=1}^{3} \gamma_{ei}\langle n|\nabla_i|m\rangle \tag{8.6.6}$$

where the γ_{ei} are direction cosines between the respective molecular axes x_i and the unit polarization vector $\mathbf{e}$. The absolute value squared of (8.6.6) is given by

$$|\langle n|\mathbf{e}\cdot\boldsymbol{\nabla}|m\rangle|^2 = \sum_{i,j=1}^{3} \gamma_{ei}\gamma_{ej}\langle n|\nabla_i|m\rangle\langle n|\nabla_i|m\rangle. \tag{8.6.7}$$

A random-orientation ensemble brings in the average value

$$\overline{\gamma_{ei}\gamma_{ej}} = \tfrac{1}{3}\delta_{ij} \tag{8.6.8}$$

by a well-known property of direction cosines. Thus

$$\overline{|\langle n|\mathbf{e}\cdot\boldsymbol{\nabla}|m\rangle|^2} = \tfrac{1}{3}\langle n|\boldsymbol{\nabla}|m\rangle\cdot\langle n|\boldsymbol{\nabla}|m\rangle^* \equiv \tfrac{1}{3}|\langle n|\boldsymbol{\nabla}|m\rangle|^2. \tag{8.6.9}$$

Note the sense of the last absolute-value bracket, a convention which will be used repeatedly in what follows. We have therefore, in place of (8.6.5), for random molecular orientations

$$W_{mn}^{\text{dipole}} = \frac{4\pi^2 e^2}{3m^2 c\omega_{nm}^2} I(\omega_{nm})|\langle n|\boldsymbol{\nabla}|m\rangle|^2. \tag{8.6.10}$$

† Alternatively, the expansion

$$e^{i\mathbf{k}\cdot\mathbf{r}} = \sum_{l=0}^{\infty} (2l+1)i^l j_l(kr)P_l(\cos\theta)$$

is useful when the wavefunctions $\psi_n(\mathbf{r})$ contain spherical harmonics.

The same result follows when it is the **e** vector which is randomized, as is the case in unpolarized isotropic radiation.

It is useful at this point to transform the matrix element of the gradient using Ehrenfest's relation for the momentum operator (cf. 4.6.5)

$$\mathbf{p} = -i\hbar\nabla = \frac{m}{i\hbar}[\mathbf{r}, \mathscr{H}^{(0)}]. \tag{8.6.11}$$

Accordingly

$$\langle n|\nabla|m\rangle = \frac{m}{\hbar^2}\langle n|(\mathbf{r}\mathscr{H}^{(0)} - \mathscr{H}^{(0)}\mathbf{r})|m\rangle. \tag{8.6.12}$$

The eigenvalue equation for $\mathscr{H}^{(0)}$ (8.5.2) implies that

$$\langle n|(\mathbf{r}\mathscr{H}^{(0)} - \mathscr{H}^{(0)}\mathbf{r}|m\rangle = (E_m - E_n)\langle n|\mathbf{r}|m\rangle = -\hbar\omega_{nm}\langle n|\mathbf{r}|m\rangle. \tag{8.6.13}$$

The transition rate can thereby be written

$$W_{mn}^{\text{dipole}} = \frac{4\pi^2 e^2}{3\hbar^2 c} I(\omega_{nm})|\langle n|\mathbf{r}|m\rangle|^2. \tag{8.6.14}$$

This is known as the *dipole approximation* because the operator

$$\boldsymbol{\mu} = -e\mathbf{r} \tag{8.6.15}$$

represents the instantaneous electric dipole moment of an electron. Introducing the *transition dipole*

$$\boldsymbol{\mu}_{nm} = \langle n|\boldsymbol{\mu}|m\rangle = -e\int \psi_n^*(\mathbf{r})\mathbf{r}\psi_m(\mathbf{r})\,\mathrm{d}^3\mathbf{r} \tag{8.6.16}$$

we can write

$$W_{mn}^{\text{dipole}} = \frac{4\pi^2}{3\hbar^2 c} I(\omega_{nm})|\boldsymbol{\mu}_{nm}|^2. \tag{8.6.17}$$

The choice of origin for the dipole operator does not affect the value of the matrix element since, for any added constant term

$$\langle n|\boldsymbol{\mu}_0|m\rangle = \boldsymbol{\mu}_0\langle n|m\rangle = 0. \tag{8.6.18}$$

Note incidentally that an electric dipole expectation value, e.g.

$$\langle\boldsymbol{\mu}\rangle_n = -e\int \psi_n^*(\mathbf{r})\mathbf{r}\psi_n(\mathbf{r})\,\mathrm{d}^3\mathbf{r} \tag{8.6.19}$$

follows from the classical formula (2.6.27) with the substitution $\rho_n = -e\psi_n^*\psi_n$. A dipole matrix element (8.6.16) can be analogously pictured as an average over a *transition density*

$$\rho_{nm}(\mathbf{r}) = -e\psi_n^*(\mathbf{r})\psi_m(\mathbf{r}) \qquad (8.6.20)$$

whereby

$$\boldsymbol{\mu}_{nm} = \int \mathbf{r}\rho_{nm}(\mathbf{r})\,\mathrm{d}^3\mathbf{r}. \qquad (8.6.21)$$

The transformation leading from (8.6.5) to (8.6.14) is exact provided that ψ_m and ψ_n are exact eigenfunctions of $\mathscr{H}^{(0)}$. Should these be approximate wavefunctions then these results will not, in general, agree. The original equation is often referred to as the "dipole velocity" form for the following reason. Observe that the Heisenberg equation of motion for the transition dipole is given by (cf. 4.5.5)

$$\dot{\boldsymbol{\mu}}_{nm} \equiv \left(\frac{\mathrm{d}\boldsymbol{\mu}}{\mathrm{d}t}\right)_{nm} = (i\hbar)^{-1}[\boldsymbol{\mu}, \mathscr{H}^{(0)}]_{nm} = i\omega_{nm}\boldsymbol{\mu}_{nm}. \qquad (8.6.22)$$

Thus (8.6.5) can be expressed

$$W_{mn}^{\text{dipole}} = \frac{4\pi^2}{3\hbar^2 c\omega_{nm}^2} I(\omega_{nm})|\dot{\boldsymbol{\mu}}_{nm}|^2 \qquad (8.6.23)$$

in which

$$\dot{\boldsymbol{\mu}}_{nm} = \frac{ie\hbar}{m}\langle n|\boldsymbol{\nabla}|m\rangle. \qquad (8.6.24)$$

One can go a step further and introduce the "dipole acceleration". In analogy with (8.6.22),

$$\ddot{\boldsymbol{\mu}}_{nm} \equiv \left(\frac{\mathrm{d}\dot{\boldsymbol{\mu}}}{\mathrm{d}t}\right)_{nm} = (i\hbar)^{-1}[\dot{\boldsymbol{\mu}}, \mathscr{H}^{(0)}]_{nm} = i\omega_{nm}\dot{\boldsymbol{\mu}}_{nm} \qquad (8.6.25)$$

so that

$$W_{mn}^{\text{dipole}} = \frac{4\pi^2}{3\hbar^2 c\omega_{nm}^2} I(\omega_{nm})|\ddot{\boldsymbol{\mu}}_{nm}|^2. \qquad (8.6.26)$$

It is interesting to note that

$$\ddot{\boldsymbol{\mu}}_{nm} + \omega_{nm}^2\boldsymbol{\mu}_{nm} = 0 \qquad (8.6.27)$$

which has the same form as the equation of motion for a classical harmonic oscillator. Computationally

$$\ddot{\boldsymbol{\mu}}_{nm} = \frac{e}{m}[\boldsymbol{\nabla}, \mathscr{H}^{(0)}] = \frac{e}{m}\langle n|\boldsymbol{\nabla}V(\mathbf{r})|m\rangle. \qquad (8.6.28)$$

Since this involves the gradient of the potential, the dipole acceleration form is generally the least accurate when applied to approximate eigenfunctions.†

The dipole transition formula (8.6.17) can alternatively be derived by considering just the effect of the electric component of the radiation field on the instantaneous dipole moment. Assuming, in addition, that there is no spatial variation in **E** over the dimensions of the system, the interaction operator is simply

$$\mathscr{V}_{\text{el dip}} = -\boldsymbol{\mu}\cdot\mathbf{E}\,. \tag{8.6.29}$$

For a polychromatic electric field

$$\mathbf{E} = \int_{-\infty}^{\infty} \mathscr{E}(\omega)\,\mathrm{e}^{-i\omega t}\,\mathrm{d}\omega \tag{8.6.30}$$

application of (8.5.17) gives

$$W_{mn} = \frac{4\pi^2}{\hbar^2 \Delta t}\,|\mathscr{E}(\omega_{nm})\cdot\boldsymbol{\mu}_{nm}|^2\,. \tag{8.6.31}$$

Assuming random orientations, as before,

$$W_{mn} = \frac{4\pi^2}{3\hbar^2 \Delta t}\,|\mathscr{E}(\omega_{nm})|^2\,|\boldsymbol{\mu}_{nm}|^2\,. \tag{8.6.32}$$

Finally, using (2.10.26) for the intensity in terms of the electric field:

$$I(\omega) = \frac{c}{\Delta t}\,|\mathscr{E}(\omega)|^2 \tag{8.6.33}$$

we arrive again at (8.7.16).

The preceding theory, although developed explicitly for a one-electron system, can be readily generalized to many electrons by interpreting the dipole operator (8.6.15) as

$$\boldsymbol{\mu} = -\mathrm{e}\sum_{i=1}^{N} \mathbf{r}_i\,. \tag{8.6.34}$$

It is instructive to rework the entire theory of Sections 8.5 and 8.6 beginning with the analogous N-electron operators

$$\mathscr{H}^{(0)} = -\frac{\hbar^2}{2m}\sum_{i=1}^{N} \nabla_i^2 + V(\mathbf{r}_1 \ldots \mathbf{r}_N) \tag{8.6.35}$$

† The three alternative forms of the transition matrix element were given by S. Chandrasekhar, *Astrophys J.* **102**, 223 (1945). For computational applications to helium, see B. Schiff and C. L. Pekeris, *Phys. Rev.* **134**, A638 (1964).

and

$$\mathscr{V} = -\frac{ie\hbar}{mc}\sum_{i=1}^{N} \mathbf{A}(\mathbf{r}_i)\cdot\nabla_i\,. \tag{8.6.36}$$

A transition between states m and n is *dipole allowed* if at least one component of the dipole matrix element $\boldsymbol{\mu}_{nm} \neq 0$. The specific conditions on the quantum numbers m and n such that $\boldsymbol{\mu}_{nm}$ does not vanish are called *dipole selection rules*. These are determined by the symmetry properties of the corresponding wavefunctions. When $\boldsymbol{\mu}_{nm} = 0$, the transition is *dipole forbidden*. Such transitions can still occur, usually with much reduced intensity, via higher multipoles—magnetic dipole, electric quadrupole, etc.—or by some perturbation mechanism which "*mixes*" the appropriate symmetry into states m or n. Suppose, for example, that the field-free Hamiltonian has the form

$$\mathscr{H}^{(0)} + \mathscr{H}^{(1)} \tag{8.6.37}$$

in which $\mathscr{H}^{(1)}$ represents some nonradiative perturbation such as vibronic coupling, spin–orbit interaction, collisions, etc. For strongly-allowed transitions, the effect of $\mathscr{H}^{(1)}$ is usually negligible. But when the dipole matrix element connecting two eigenfunctions $\psi_n^{(0)}$ and $\psi_m^{(0)}$ of $\mathscr{H}^{(0)}$ vanishes, i.e.,

$$\boldsymbol{\mu}_{nm}^{(0)} \equiv \int \psi_n^{(0)*}\boldsymbol{\mu}\psi_n^{(0)}\,d\tau = 0 \tag{8.6.38}$$

then the influence of $\mathscr{H}^{(1)}$ must be taken into account. By time-independent perturbation theory, the perturbed states ψ_n and ψ_m are represented, to first order, by

$$\psi_n^* = \psi_n^{(0)*} - \mathop{\mathbf{S}'}_{k} \frac{H_{nk}^{(1)}\psi_k^{(0)*}}{E_k^{(0)} - E_n^{(0)}}, \quad \psi_m = \psi_m^{(0)} - \mathop{\mathbf{S}'}_{k} \frac{\psi_k^{(0)}H_{km}^{(1)}}{E_k^{(0)} - E_m^{(0)}}\,. \tag{8.6.39}$$

Therefore

$$\boldsymbol{\mu}_{nm} \equiv \mathbf{S}\psi_n^*\boldsymbol{\mu}\psi_m\,dr = \boldsymbol{\mu}_{nm}^{(0)} - \mathop{\mathbf{S}'}_{k}\left[\frac{H_{nk}^{(1)}\boldsymbol{\mu}_{km}^{(0)}}{E_k^{(0)} - E_k^{(0)}} + \frac{\boldsymbol{\mu}_{nk}^{(0)}H_{km}^{(1)}}{E_k^{(0)} - E_m^{(0)}}\right]. \tag{8.6.40}$$

This will be non-vanishing even if $\boldsymbol{\mu}_{nm}^{(0)} = 0$ provided there exists a set of states k with the appropriate nonzero matrix elements of $\boldsymbol{\mu}$ and $\mathscr{H}^{(1)}$. For example, in the electronic spectra of molecules, certain symmetry-forbidden transitions are attributed to vibronic coupling, which mixes the appropriate symmetry-allowed character into the upper and lower states (Herzberg–Teller effect). Again, certain spin-forbidden (e.g. singlet → triplet) transitions do occur weakly because of spin–orbit coupling.

8.7 Oscillator Strength

A convenient standard of intensity for electronic transitions in atoms and molecules can be taken as the ratio

$$f_{mn} \equiv \frac{W_{mn}}{W_{mn}^{\mathrm{HO}}} \tag{8.7.1}$$

known as the *oscillator strength* or *f-number*. Here W_{mn}^{HO} represents the rate of the $0 \to 1$ transition in a spherical harmonic oscillator of fundamental frequency $\omega_0 = \omega_{nm}$. For a given radiation intensity,

$$f_{mn} = \frac{|\boldsymbol{\mu}_{nm}|^2}{e^2|\langle 0|\mathbf{r}|1\rangle|^2}\,. \tag{8.7.2}$$

The appropriate spherical harmonic oscillator matrix elements are

$$x_{01} = y_{01} = z_{01} = (\hbar/2m\omega_{nm})^{\frac{1}{2}} \tag{8.7.3}$$

so that†

$$f_{mn} = \frac{2m\omega_{nm}}{3\hbar e^2}|\boldsymbol{\mu}_{nm}|^2\,. \tag{8.7.4}$$

Note that the oscillator strength, as defined, has *negative* values for downward transitions ($\omega_{nm} < 0$). Moreover

$$f_{nm} = -f_{mn}\,. \tag{8.7.5}$$

In terms of oscillator strength, the transition rate can be written

$$W_{mn} = \frac{2\pi^2 e^2}{mc\hbar\omega_{mn}} I(\omega_{nm}) f_{mn}\,. \tag{8.7.6}$$

A harmonically-bound electron in classical theory has $f = 1$. Thus the classical formula for absorption of power from a radiation field is

$$P = \hbar\omega W = \frac{2\pi^2 e^2}{mc} I(\omega) \tag{8.7.7}$$

† Oscillator strength was actually first introduced in the theory of dispersion. Electric polarizability as a function of frequency can be represented in the form

$$\alpha(\omega) = \frac{e^2}{m} \mathop{\mathbf{S}}_{n} \frac{f_{mn}}{\omega_{nm}^2 - \omega^2}\,.$$

The corresponding classical result for a single harmonically-bound electron is

$$\alpha(\omega) = \frac{e^2}{m(\omega_0^2 - \omega^2)}$$

equivalent to unit oscillator strength. The quantum result can hence be conceptualized as arising from a series of oscillators of frequency ω_{nm} and relative strength f_{nm}.

More completely,

$$P = \frac{2\pi^2 e^2}{mc} I(\omega) \cos \delta \tag{8.7.8}$$

where δ is the phase difference between the radiation and the oscillator. Maximum absorption occurs when the two are precisely in phase, $\delta = 0$. Maximum stimulated emission occurs when the field and oscillator are exactly out of phase, $\delta = \pi$. Just as in quantum theory, these two rates are equal. By contrast, however, classical theory predicts a continuum between absorption and stimulated emission, depending on the phase difference.

Oscillator strengths satisfy a very interesting sum rule. Let us sum f_{mn}, eqn (8.7.4), over all possible final states n. Using the hermitian property of the matrix element,

$$\mathop{\mathbf{S}}_n f_{mn} = \frac{2m}{3\hbar^2} \mathop{\mathbf{S}}_n (E_n - E_m)\mathbf{r}_{nm} \cdot \mathbf{r}_{nm} \,. \tag{8.7.9}$$

But by (8.6.13) and (8.6.11)

$$(E_n - E_m)\mathbf{r}_{mn} = [\mathbf{r}, \mathscr{H}^{(0)}]_{mn} = \frac{i\hbar}{m} \mathbf{p}_{mn} \,. \tag{8.7.10}$$

Also

$$(E_n - E_m)\mathbf{r}_{nm} = [\mathscr{H}^{(0)}, \mathbf{r}]_{nm} = -\frac{i\hbar}{m} \mathbf{p}_{nm} . \tag{8.7.11}$$

Thus (8.7.9) can be written

$$\mathop{\mathbf{S}}_n f_{mn} = \frac{i}{3\hbar} \mathop{\mathbf{S}}_n (\mathbf{p}_{mn} \cdot \mathbf{r}_{nm} - \mathbf{r}_{mn} \cdot \mathbf{p}_{nm}) \,. \tag{8.7.12}$$

By matrix multiplication and the fundamental commutation relations (3.4.4)

$$\mathop{\mathbf{S}}_n (\mathbf{p}_{mn} \cdot \mathbf{r}_{nm} - \mathbf{r}_{mn} \cdot \mathbf{p}_{nm}) = \langle n|\mathbf{p} \cdot \mathbf{r} - \mathbf{r} \cdot \mathbf{p}|m\rangle$$

$$= \sum_{i=1}^{3} [p_i, x_i]_{mn} = -3i\hbar \,. \tag{8.7.13}$$

Therefore

$$\mathop{\mathbf{S}}_n f_{mn} = 1. \tag{8.7.14}$$

This is the famous Kuhn–Thomas sum rule. It was originally derived before quantum mechanics on the basis of classical dispersion theory and was, in fact, what led Heisenberg to postulate the fundamental commutation relations in matrix mechanics.

The sum rule (8.7.14) applies rigorously to one-electron systems or approximately to N-electron systems for the set of one-electron excitations from a given state. This can be generalized to the case of N electrons by using the Hamiltonian (8.6.35) and the dipole operator (8.6.34). The result, taking into account all possible transitions, is

$$\mathop{\mathbf{S}}_{n} f_{mn} = N. \tag{8.7.15}$$

For transitions from the ground state ($m = 0$) in the one-electron case, we must have $0 \leqslant f_{on} \leqslant 1$. For strongly allowed transitions, $f_{on} \sim 1$, evidently. For example, the yellow sodium D-line (actually a doublet $^2S_{\frac{1}{2}} \rightarrow {}^2P_{\frac{1}{2}}, {}^2P_{\frac{3}{2}}$), has an oscillator strength of 0·976. For a large class of organic compounds, dipole-allowed transitions have typically $f \sim 0{\cdot}1$ to 1, symmetry-forbidden transitions via vibronic mechanisms, $f \sim 10^{-4}$ to 10^{-2} and spin-forbidden transitions via spin–orbit coupling may have $f \sim 10^{-5}$.

8.8 Spontaneous Emission

Excited atoms and molecules are known to emit radiation even in the absence of external fields. This *spontaneous emission* cannot be rigorously treated within the semiclassical theory of radiation we have been using. The process should not even occur, in fact, since in the absence of external perturbations, stationary states—even excited ones—ought to be maintained indefinitely. The quantum theory of radiation, begun by Dirac in 1927, successfully explained spontaneous emission, while reproducing the semiclassical results on absorption and stimulated emission. According to quantum electrodynamics, even the vacuum state of the radiation field exhibits a finite interaction with matter, so that no excited states of atoms or molecules can be strictly stationary.

It is possible nonetheless to give a fairly plausible account of spontaneous emission on the basis of semiclassical theory. We recall the formula (2.9.26) for the power radiated by a classical oscillator of frequency ω

$$P(\omega) = \frac{4\omega^4}{3c^3}\,|\boldsymbol{\mu}(\omega)|^2 \tag{8.8.1}$$

where $\boldsymbol{\mu}(\omega)$ is the Hertzian dipole vector (cf. 2.9.12). It is intuitively reasonable to associate (8.8.1) with the energy per unit time emitted in spontaneous downward transitions $n \rightarrow m$, with $\omega = \omega_{nm}$. Since the quantized energy of each such transition is $\hbar\omega_{nm}$, the spontaneous transition rate (probability per unit time) is evidently given by

$$W_{nm}^{\text{sp em}} = \frac{P(\omega_{nm})}{\hbar\omega_{nm}}. \tag{8.8.2}$$

The *ad hoc* postulate is now made that the classical radiation formula (8.8.1) becomes applicable in quantum theory if the Hertzian dipole is replaced by an appropriate dipole matrix element. To determine this correspondence precisely, we examine the relation (2.9.15) with $\omega = \omega_{nm}$, i.e.

$$\boldsymbol{\mu}(\omega_{nm}) = \int \mathbf{r}\rho_{\omega_{nm}}(\mathbf{r})\,\mathrm{d}^3\mathbf{r} \tag{8.8.3}$$

which gives the Hertzian dipole in terms of a fourier amplitude of oscillating charge density. Now $\rho_{\omega_{nm}}(\mathbf{r})$ is associated with a time factor $\mathrm{e}^{-i\omega_{nm}t}$. The product of eigenfunctions $\psi_m^*(\mathbf{r})\,\psi_m(\mathbf{r})$ has the very same time factor in its time-dependent form. This suggests the quantization prescription

$$\rho_{\omega_{nm}} \to -e\psi_n\psi_m^* \tag{8.8.4}$$

the latter quantity being the transition density $\rho_{mn}(\mathbf{r})$ defined in eqn (8.6.20). In terms of the matrix element (8.6.16)

$$\boldsymbol{\mu}_{mn} = -e\int \psi_m^*(\mathbf{r})\mathbf{r}\psi_n(\mathbf{r})\,\mathrm{d}^3\mathbf{r} \tag{8.8.5}$$

(8.8.4) is equivalent to

$$\boldsymbol{\mu}(\omega_{nm}) \to \boldsymbol{\mu}_{mn}\,. \tag{8.8.6}$$

The plausibility of this correspondence is further enhanced by the complex-conjugation property of the Hertzian dipole (cf. 2.9.13):

$$\boldsymbol{\mu}^*(\omega_{nm}) = \boldsymbol{\mu}(-\omega_{nm}) = \boldsymbol{\mu}(\omega_{mn}) \tag{8.8.7}$$

which matches the hermitian property of the matrix element:

$$\boldsymbol{\mu}_{mn}^* = \boldsymbol{\mu}_{mn}\,. \tag{8.8.8}$$

Putting (8.8.2) and (8.8.6) into (8.8.1), we obtain the quantum formula for electric-dipole spontaneous emission:

$$W_{nm}^{\text{sp em}} = \frac{4\omega_{nm}^3}{3\hbar c^3}|\boldsymbol{\mu}_{nm}|^2\,. \tag{8.8.9}$$

Comparing this with the rate of the corresponding stimulated emission process, eqn (8.6.17), we find the ratio

$$\frac{W_{nm}^{\text{sp em}}}{W_{nm}^{\text{stim em}}} = \frac{\hbar\omega_{nm}^3}{\pi^2c^2I(\omega_{nm})}\,. \tag{8.8.10}$$

It can therefore be postulated that a more general formula is given by the same multiple of (8.5.19):

$$W_{nm}^{\text{sp em}} = \frac{4\hbar e^2\omega_{nm}}{m^2c^3}|\langle n|\mathrm{e}^{i\mathbf{k}\cdot\mathbf{r}}\,\mathbf{e}\cdot\nabla|m\rangle|^2\,. \tag{8.8.11}$$

This pertains to the rate of emission of photons with wavevector **k** and polarization **e**. Note that this result requires neither the dipole approximation nor spherical averaging. It is, however, limited to single-photon processes (first-order perturbation theory). Since spontaneous emission depends on the same matrix elements as do the radiation-induced processes, the very same selection rules will pertain.

The spontaneous transition rate (8.8.9) can evidently be identified with the parameter γ in Section 4.12. More precisely, if γ represents the *total* decay rate of the state ψ_n, then

$$\gamma = \sum_{\substack{m \\ (E_m < E_n)}} W_{nm}^{\text{sp em}} \tag{8.8.12}$$

which takes into account the possibility of spontaneous transitions to all states of lower energy. As discussed in Section 4.12, γ approximates the natural linewidth of spectral lines involving the state ψ_n. The average lifetime of ψ_n is (cf. 4.12.5)

$$\Delta t = 1/\gamma \tag{8.8.13}$$

which reduces to the reciprocal of $W_{nm}^{\text{sp em}}$ if there is just one lower state m. For dipole-allowed electronic transitions, Δt is typically of the order of 10^{-8} sec. This magnitude can be estimated from (8.8.9) by setting $\omega \sim 5 \times 10^{15}$ sec^{-1} and $\mu \sim 1D \sim 5 \times 10^{-18}$ esu-cm.

The rate of spontaneous emission is very strongly dependent on frequency owing to the ω^3 factor. Hence for microwave and radiofrequencies ($\omega \gtrsim 10^{12}$ sec^{-1}), the process is entirely negligible.† For nuclear γ-rays, on the other hand ($\omega \sim 10^{19}$ to 10^{21} sec^{-1}), spontaneous emission becomes the dominant radiative process.

8.9 Blackbody Radiation

The hypothetical perfect absorber of all wavelengths of electromagnetic radiation is known as a *blackbody*. The best laboratory approximation is a large insulated cavity (*hohlraum*) in internal thermal equilibrium. The blackbody (or thermal) radiation inside the cavity can be observed through a small orifice. The spectral distribution of the radiation can be characterized by its energy density as a function of frequency, $\rho(\omega)$. Since radiation travels at a speed c, the energy density is related to the intensity by

$$I(\omega) = c\rho(\omega)\,. \tag{8.9.1}$$

Since the radiation is isotropic, its description in terms of $\rho(\omega)$ is somewhat more appropriate. It is observed that $\rho(\omega)$ is determined by the temperature

† A notable exception is the 21-cm spin–flip transition in atomic hydrogen, observed in radio astronomy. The average lifetime of a spin-excited hydrogen atom is of the order of 1000 years.

alone. The spectrum is continuous with a maximum in intensity at a wavelength inversely proportional to temperature, in accordance with Wien's displacement law:

$$\lambda_{max} = \frac{0{\cdot}2896}{T}. \tag{8.9.2}$$

At room temperature, the maximum lies in the infrared. As T is increased above 1000 K, λ_{max} progresses through the visible region: a body becomes successively red-hot, white-hot and eventually blue-hot. The integrated blackbody energy density increases very sharply with temperature, as described by the Stefan–Boltzmann law:

$$\rho_{total} = \int_0^{\infty} \rho(\omega)\, d\omega = 7{\cdot}563 \times 10^{-15}\, T^4. \tag{8.9.3}$$

The blackbody distribution law can be derived from the preceding results on absorption and emission of radiation. Suppose that the molecules of the cavity walls have attained equilibrium with the radiation field contained therein. Consider a pair of molecular energy levels E_n and E_m ($E_n > E_m$), between which transitions are allowed. The rate of absorption of photons of energy $\hbar\omega_{nm}$, wavevector $\mathbf{k}$ and polarization $\mathbf{e}$ is equal to $N_m W_{mn}^{abs}$, where N_m is the population of the state m and W_{mn}^{abs} is given by the general expression (8.5.19). The rate of emission of the same photons is analogously $N_n(W_{nm}^{stim\,em} + W_{nm}^{sp\,em})$, taking into consideration the two emission mechanisms. In thermal equilibrium, the rates of each elementary process and its inverse are equal (principle of detailed balancing). Thus for each pair of levels n, m:

$$N_m W_{mn}^{abs} = N_n(W_{nm}^{stim\,em} + W_{nm}^{sp\,em}). \tag{8.9.4}$$

Assuming a Boltzmann distribution, the molecular populations are in the ratio

$$\frac{N_m}{N_n} = e^{\hbar\omega_{nm}/kT}. \tag{8.9.5}$$

We now make use of the equality of W_{mn}^{abs} *and* $W_{nm}^{stim\,em}$ and of the ratio (8.8.10) between $W_{nm}^{sp\,em}$ and $W_{nm}^{stim\,em}$. We obtain thereby

$$e^{\hbar\omega/kT} = 1 + \frac{\hbar\omega^3}{\pi^2 c^2 I(\omega)} \tag{8.9.6}$$

the subscripts on ω having been dropped since the result pertains to arbitrary pairs of energy levels. Solving for the energy density (cf. 8.9.1), there results

$$\rho(\omega) = \frac{\hbar\omega^3/\pi^2 c^3}{e^{\hbar\omega/kT} - 1}. \tag{8.9.7}$$

This is the famous Planck distribution formula which was a precursor of the quantum theory. It is more customarily expressed in terms of the frequency ν as

$$\rho(\nu) = \frac{8\pi h\nu^3/c^3}{e^{h\nu/kT} - 1}. \tag{8.9.8}$$

It is interesting to note that all parameters characteristic of particular molecular species have dropped out—as indeed have even the constants e and m. The fact that we have obtained Planck's radiation law correctly also provides a final verification of the spontaneous emission formulas (8.8.9) and (8.8.11).†

One can also express the blackbody distribution formula in terms of wavelength as follows

$$\rho(\lambda) = \frac{8\pi hc/\lambda^5}{e^{hc/\lambda kT} - 1}. \tag{8.9.9}$$

Wien's displacement law (8.9.2) pertains to the maximum value of (8.9.9). Setting $\rho'(\lambda) = 0$ leads to the transcendental equation

$$x = 5(1 - e^{-x}), \; x \equiv hc/\lambda_{max} kT. \tag{8.9.10}$$

The numerical solution is $x = 4{\cdot}96511$. Thus

$$\lambda_{max} = \frac{hc/k}{4{\cdot}96511\,T} = \frac{0{\cdot}289780}{T}. \tag{8.9.11}$$

The Stefan–Boltzmann law (8.9.3) is obtained by integration of, say, (8.9.8):

$$\begin{aligned}\rho_{total} &= \int_0^\infty \rho(\nu)\,d\nu = \frac{8\pi h}{c^3}\int_0^\infty \frac{\nu^3\,d\nu}{e^{h\nu/kT} - 1} \\ &= \frac{8\pi k^4 T^4}{h^3 c^3}\int_0^\infty \frac{x^3\,dx}{e^x - 1}\end{aligned} \tag{8.9.12}$$

The definite integral has the value $\pi^4/15$, so that

$$\rho_{total} = 7{\cdot}56493 \times 10^{-15}\,T^4. \tag{8.9.13}$$

8.10 Higher Multipole Transitions

Even when a transition $m \to n$ is electric-dipole forbidden, i.e.,

$$\boldsymbol{\mu}_{nm} = 0 \tag{8.10.1}$$

† Einstein (1917) first deduced the ratio of spontaneous to stimulated emission by applying the above phenomenological argument starting with Planck's radiation law. Our derivation has, in fact, reversed the historical input and output.

it might still occur with weaker intensity if the second term of the series (8.6.4) gives rise to a nonvanishing matrix element. Assuming (8.10.1), the general formula for radiation-induced transitions, eqn (8.5.19), has the leading term

$$W_{mn} = \frac{4\pi^2 e^2}{m^2 c \omega_{nm}^2} I(\omega_{nm}) |\langle n|\mathbf{k}\cdot\mathbf{r}\mathbf{e}\cdot\nabla|m\rangle|^2. \tag{8.10.2}$$

Recall that

$$\mathbf{k} = k\mathbf{n} = \frac{\omega_{nm}}{c}\mathbf{n} \tag{8.10.3}$$

in which **n** is the unit vector in the propagation direction. Thus

$$W_{mn} = \frac{4\pi^2 e^2}{m^2 c^3} I(\omega_{nm}) |\langle n|\mathbf{n}\cdot\mathbf{r}\mathbf{e}\cdot\nabla|m\rangle|^2. \tag{8.10.4}$$

Referring back to (8.5.9), the original radiation operator, shows that one need not be concerned with the action of ∇ except on the wavefunction. We now write the operator product in the matrix element as a sum of symmetric and antisymmetric forms:

$$\mathbf{n}\cdot\mathbf{r}\mathbf{e}\cdot\nabla = \tfrac{1}{2}(\mathbf{n}\cdot\mathbf{r}\mathbf{e}\cdot\nabla + \mathbf{e}\cdot\mathbf{r}\mathbf{n}\cdot\nabla) + \tfrac{1}{2}(\mathbf{n}\cdot\mathbf{r}\mathbf{e}\cdot\nabla - \mathbf{e}\cdot\mathbf{r}\mathbf{n}\cdot\nabla) \tag{8.10.5}$$

The vector identity

$$\mathbf{n}\times\mathbf{e}\cdot\mathbf{r}\times\nabla = \mathbf{n}\cdot\mathbf{r}\mathbf{e}\cdot\nabla - \mathbf{e}\cdot\mathbf{r}\mathbf{n}\cdot\nabla \tag{8.10.6}$$

pertains to the antisymmetric part. We introduce the angular-momentum operator

$$\mathscr{L} = -i\hbar\mathbf{r}\times\nabla \tag{8.10.7}$$

and write

$$\mathbf{n}\times\mathbf{e} \equiv \mathbf{e}' \tag{8.10.8}$$

for the unit vector normal to both **n** and **e**, in the direction of the magnetic field. The antisymmetric part of (8.10.5) thus gives rise to the transition rate

$$W_{mn}^{\text{mag dipole}} = \frac{4\pi^2 e^2}{m^2 \hbar^2 c^3} I(\omega_{nm}) |\langle n|\tfrac{1}{2}\mathbf{e}'\cdot\mathscr{L}|m\rangle|^2. \tag{8.10.9}$$

Clearly, it is indicated to define the magnetic dipole operator (cf. 2.6.45):

$$\mathbf{m} \equiv \frac{e}{2mc}\mathscr{L}. \tag{8.10.10}$$

For random molecular orientations or for isotropic radiation, we obtain

$$W_{mn}^{\text{mag dipole}} = \frac{4\pi^2}{3\hbar^2 c} I(\omega_{nm}) |\mathbf{m}_{nm}|^2 \tag{8.10.11}$$

in precise analogy with the electric-dipole formula (8.6.17). This result can be made applicable as well to transitions involving electron spins by generalizing the magnetic dipole operator to

$$\mathbf{m} = \frac{e}{2mc}(\mathscr{L} + 2\mathscr{S}) \tag{8.10.12}$$

in which $\mathscr{S}$ is the spin angular momentum, with its associated g-factor approximated by 2.

The relative intensities of electric and magnetic dipole transitions in atoms and molecules can be inferred from the typical magnitudes of their respective dipole matrix elements:

$$\frac{m}{\mu} \sim \frac{e\hbar/2mc}{ea_0} = \frac{e^2}{2\hbar c} = \frac{\alpha}{2}. \tag{8.10.13}$$

Thus

$$\frac{W^{\text{mag dipole}}}{W^{\text{elec dipole}}} \sim \frac{\alpha^2}{4} \sim 10^{-5}. \tag{8.10.14}$$

Consequently, one never worries about magnetic dipole contributions in electric-dipole allowed transitions. (However, optical activity involves cross-terms between electric and magnetic matrix elements.)

In analogy with the calculation of eqns (8.6.29–33), the magnetic dipole result (8.10.12) also follows from the interaction of a magnetic dipole with the magnetic component of the radiation field:

$$\mathscr{V}_{\text{mag}} = -\,\mathbf{m}\cdot\mathbf{H}, \quad \mathbf{H} = \int_{-\infty}^{\infty} \mathscr{H}(\omega)\,\mathrm{e}^{-i\omega t}\,\mathrm{d}\omega. \tag{8.10.15}$$

We turn now to the symmetric term in the identity (8.10.5). Transforming the momentum operator as in (8.6.11), we have

$$\begin{aligned}(\mathbf{n}\cdot\mathbf{r}\,\mathbf{e}\cdot\nabla + \mathbf{e}\cdot\mathbf{r}\,\mathbf{n}\cdot\nabla) &= \frac{m}{\hbar^2}(\mathbf{n}\cdot\mathbf{r},[\mathbf{e}\cdot\mathbf{r},\mathscr{H}^{(0)}] + \mathbf{e}\cdot\mathbf{r}[\mathbf{n}\cdot\mathbf{r},\mathscr{H}^{(0)}]) \\ &= \frac{m}{\hbar^2}[\mathbf{n}\cdot\mathbf{r}\,\mathbf{e}\cdot\mathbf{r},\mathscr{H}^{(0)}] = \frac{m}{\hbar^2}\mathbf{n}\cdot[\mathbf{rr},\mathscr{H}^{(0)}]\cdot\mathbf{e}\end{aligned} \tag{8.10.16}$$

We have used the commutator identity (3.3.18) coupled with the fact that **n** and **e** are orthogonal (so that $\mathbf{e}\cdot\mathbf{r}$ and $\mathbf{n}\cdot\nabla$ commute). In analogy with (8.6.13), the matrix element of the last commutator is given by

$$\langle n|[\mathbf{rr},\mathscr{H}^{(0)}]|m\rangle = \frac{\hbar\omega_{nm}}{e}\mathfrak{Q}'_{nm} \tag{8.10.17}$$

having introduced the *electric quadrupole operator* (cf. 2.6.28)

$$\mathfrak{Q}' \equiv -e\mathbf{rr}\,. \tag{8.10.18}$$

Since **n** and **e** are orthogonal, we can equally well use the traceless form of the quadrupole tensor (cf. 2.6.32):

$$\mathfrak{Q} \equiv -e(3\mathbf{rr} - r^2\mathfrak{J}). \tag{8.10.19}$$

We obtain thus for the electric quadrupole transition rate (assuming the absence of both electric and magnetic dipole contributions):

$$W_{mn}^{\text{el quad}} = \frac{\pi^2 \omega_{nm}^2}{9\hbar^2 c^3} I(\omega_{nm}) |\mathbf{n} \cdot \mathfrak{Q}_{nm} \cdot \mathbf{e}|^2. \tag{8.10.20}$$

Averaging (8.10.20) over molecular orientations (or for isotropic radiation is somewhat more complicated than in the dipole case. The calculation is lightened somewhat by referring the tensor $\mathfrak{Q}_{nm}$ to its principal axes (cf. 2.6.40):

$$\mathfrak{Q}_{nm} = \mathbf{ii}Q_1 + \mathbf{jj}Q_2 + \mathbf{kk}Q_3\,. \tag{8.10.21}$$

(Do not confound cartesian direction indices i, j with matrix indices m, n.) Then, in terms of the direction cosines of **n** and **e** referred to these axes

$$\mathbf{n} \cdot \mathfrak{Q}_{nm} \cdot \mathbf{e} = \sum_{i=1}^{3} Q_i \gamma_{ni} \gamma_{ei} \tag{8.10.22}$$

and

$$\begin{aligned} |\mathbf{n} \cdot \mathfrak{Q}_{nm} \cdot \mathbf{e}|^2 &= \sum_{i,j=1}^{3} Q_i Q_j^* \gamma_{ni} \gamma_{nj} \gamma_{ei} \gamma_{ej} \\ &= \sum_{i=1}^{3} |Q_i|^2 \gamma_{ni}^2 \gamma_{ei}^2 + \sum_{i,j=1}^{3}{}' Q_i Q_j^* \gamma_{ni} \gamma_{nj} \gamma_{ei} \gamma_{ej}\,. \end{aligned} \tag{8.10.23}$$

The prime on the last double summation signifies exclusion of diagonal terms. We now make use of two averages over direction cosines:†

† If $\gamma_{ni} \equiv \cos\theta$ then γ_{ei} is $\sin\theta\cos\phi$ for **e** normal to **n**. Thus

$$\overline{\gamma_{ni}^2 \gamma_{ei}^2} = \overline{\cos^2\theta \sin^2\theta \cos^2\phi} = \tfrac{1}{15}\,.$$

The second formula follows from the orthogonality condition

$$\sum_{i=1}^{3} \gamma_{ni} \gamma_{ei} = 0\,.$$

Squaring

$$\sum_{ij=1}^{3} \gamma_{ni} \gamma_{nj} \gamma_{ei} \gamma_{ej} = \sum_{i=1}^{3} \gamma_{ni}^2 \gamma_{ei}^2 + \sum_{i,j=1}^{3} \gamma_{ni} \gamma_{nj} \gamma_{ei} \gamma_{ej} = 0.$$

Averaging over orientations gives

$$3\overline{\gamma_{ni}^2 \gamma_{ei}^2} + 6\overline{\gamma_{ni} \gamma_{nj} \gamma_{ei} \gamma_{ej}} = 0$$

whence (8.10.25) follows from (8.10.24).

$$\overline{\gamma_{ni}^2 \gamma_{ei}^2} = \tfrac{1}{15} \tag{8.10.24}$$

$$\overline{\gamma_{ni}\gamma_{nj}\gamma_{ei}\gamma_{ej}} = -\tfrac{1}{30} \quad (i \neq j). \tag{8.10.25}$$

These imply that

$$\overline{|\mathbf{n}\cdot\mathfrak{Q}_{nm}\cdot\mathbf{e}|^2} = \tfrac{1}{15}\sum_{i=1}^{3}|Q_i|^2 - \tfrac{1}{30}\sum_{i,j=1}^{3}{}' \, Q_i Q_j^*. \tag{8.10.26}$$

Since $\mathfrak{Q}$ is traceless

$$\sum_{i=1}^{3} Q_i = 0, \quad \sum_{i,j=1}^{3} Q_i Q_j^* = \sum_{i=1}^{3}|Q_i|^2 + \sum_{i,j=1}^{3}{}' \, Q_i Q_j^* = 0 \tag{8.10.27}$$

whereby

$$\overline{|\mathbf{n}\cdot\mathfrak{Q}_{nm}\cdot\mathbf{e}|^2} = \tfrac{1}{10}\sum_{i=1}^{3}|Q_i|^2. \tag{8.10.28}$$

For arbitrary axes, the last summation is equivalent to

$$\sum_{i,j=1}^{3} Q_{ij}Q_{ij}^* = \mathfrak{Q}_{nm} : \mathfrak{Q}_{nm}^* \equiv |\mathfrak{Q}_{nm}|^2. \tag{8.10.29}$$

Thus the ensemble-averaged quadrupole radiation rate is given by

$$W_{mn}^{\text{el quad}} = \frac{\pi^2 \omega_{nm}}{90\hbar c^3} I(\omega)_{nm} |\mathfrak{Q}_{nm}|^2. \tag{8.10.30}$$

A typical non-vanishing quadrupole matrix element has the order of magnitude

$$Q \sim eA^2 \tag{8.10.31}$$

($A = 10^{-8}$ cm). Thus the ratio of electric–quadrupole to electric–dipole intensities will be of the order

$$\frac{W^{\text{el quad}}}{W^{\text{el dip}}} \sim \frac{\omega^2 A^2}{120 c^2} \sim 10^{-8} \tag{8.10.32}$$

for optical transitions ($\omega \sim 5 \times 10^{15}$ sec^{-1}).

Again, the quadrupole radiation formula can alternatively be derived by considering just the appropriate part of the radiative perturbation (cf. (2.6.34))

$$\mathscr{V}_{\text{el quad}} = -\tfrac{1}{6}\mathfrak{Q} : \nabla\mathbf{E}. \tag{8.10.33}$$

By (2.10.18)

$$\mathbf{E} = i\int_{-\infty}^{\infty} k\mathscr{A}(\omega)\, e^{i(\mathbf{k}\cdot\mathbf{r}-\omega t)}\, d\omega. \tag{8.10.34}$$

Thus

$$\nabla\mathbf{E} = -\int_{-\infty}^{\infty} k\mathbf{k}\mathscr{A}(\omega)\,\mathrm{e}^{i(\mathbf{k}\cdot\mathbf{r}-\omega t)}\,\mathrm{d}\omega$$

$$\approx -\int_{-\infty}^{\infty} k\mathbf{k}\mathscr{A}(\omega)\,\mathrm{e}^{-i\omega t}\,\mathrm{d}\omega \tag{8.10.35}$$

A Fourier component of (8.10.33) has the form

$$\mathscr{V}(\omega) = -\frac{\omega^2}{6c^2}\mathbf{n}\cdot\mathfrak{Q}\cdot\mathscr{A}(\omega). \tag{8.10.36}$$

By (8.5.17) and (8.5.18) we obtain the transition rate

$$W_{mn}^{\text{el quad}} = \frac{4\pi^2}{\hbar^2\Delta t}\left(\frac{\omega^2}{6c^2}\right)^2 |\mathscr{A}(\omega)|^2|\mathbf{n}\cdot\mathfrak{Q}_{nm}\cdot\mathbf{e}|^2$$

$$= \frac{\pi^2\omega_{nm}^2}{9\hbar^2c^3}I(\omega_{nm})|\mathbf{n}\cdot\mathfrak{Q}_{nm}\cdot\mathbf{e}|^2 \tag{8.10.37}$$

in agreement with (8.10.20).

Appendix A
The Dirac Deltafunction

Dirac's innovation has provided an extremely valuable adjunct to the mathematical formulation of quantum mechanics. In its history as a mathematical object, the deltafunction has undergone the gamut from ridicule to commonplace acceptance. Thus, Professor Lighthill dedicates his book† as follows:

> To Paul Dirac, who saw that it must be true, Laurent Schwartz, who proved it, and George Temple, who showed how simple it could be made.

The deltafunction has the following definitive property:

$$\int_{-\infty}^{\infty} f(x)\delta(x-a)\,\mathrm{d}x = f(a), \tag{A.1}$$

where $f(x)$ is an arbitrary analytic function. By an appropriate shift of origin, this can be expressed

$$\int_{-\infty}^{\infty} f(x)\delta(x)\,\mathrm{d}x = f(0). \tag{A.2}$$

One picture of the deltafunction consistent with (A.2) might be the following: (i) it is equal to zero for all values of x except those in an infinitesimal segment around $x = 0$; (ii) within that segment, it has an infinite value, i.e.

$$\delta(x) = \begin{cases} 0\,, & x \neq 0 \\ \infty, & x = 0 \end{cases} \tag{A.3}$$

(iii) the infinite singularity is, however, compatible with the normalization condition

$$\int_{-\infty}^{\infty} \delta(x)\,\mathrm{d}x = 1\,. \tag{A.4}$$

† M. J. Lighthill, "Fourier Analysis and Generalised Functions", Cambridge University Press, 1958, containing an excellent account of the deltafunction from a mathematically rigorous point of view.

By virtue of (A.3), the range of integration in (A.1) or (A.2) need only enclose the singular point. When the singular point is one of the limits, it contributes half its value, for example.

$$\int_{a}^{\infty} f(x)\delta(x - a)\,\mathrm{d}x = \tfrac{1}{2}f(a). \tag{A.5}$$

The deltafunction is evidently the continuum analog of the Kronecker delta:

$$\delta_{n,m} = \begin{cases} 1 & \text{if } m = n \\ 0 & \text{if } m \neq n. \end{cases} \tag{A.6}$$

With respect to an arbitrary function f_n of a discrete variable n, the Kronecker delta has the following summation property:

$$\sum_{n} f_n \delta_{n,m} = f_m. \tag{A.7}$$

In particular, for $f_n = 1$ (all n),

$$\sum_{n} \delta_{n,m} = 1. \tag{A.8}$$

Equations (A.1) and (A.4) are obviously the continuum analogs of (A.7) and (A.8), respectively.

In view of its extraordinarily singular character, the deltafunction does not fit into the usual categories of mathematical functions. It can, however, be accommodated within the domain of *generalized functions*. A generalized function is defined as the limit, under appropriate conditions, of a sequence of analytic, integrable functions. The deltafunction, for example, can be realized in one possible way by a limiting process in which a normalized gaussian distribution is continuously deformed into an infinitely-sharp, infinitesimally-narrow peak. Explicitly,

$$\delta(x) = \lim_{\sigma \to 0} \frac{1}{\sqrt{(2\pi)}\sigma} \mathrm{e}^{-x^2/2\sigma^2}. \tag{A.9}$$

As $\sigma \to 0$, the function tends to infinity at $x = 0$ while approaching zero elsewhere, in accord with (A.3). Furthermore, the normalization condition (A.4) is exactly fulfilled for each value of σ since

$$\frac{1}{\sqrt{(2\pi)}\sigma} \int_{-\infty}^{\infty} \mathrm{e}^{-x^2/2\sigma^2}\,\mathrm{d}x = 1. \tag{A.10}$$

An analogous representation can be based on the normalized Lorentzian distribution function, viz,

$$\delta(x) = \lim_{a \to 0} \frac{a/\pi}{x^2 + a^2}. \tag{A.11}$$

Two further representations of the deltafunction have utility in specific applications. Consider first

$$\delta(x) = \lim_{k\to\infty} \frac{\sin kx}{\pi x}. \tag{A.12}$$

By virtue of the well-known definite integral

$$\int_{-\infty}^{\infty} \frac{\sin x}{x}\, \mathrm{d}x = \pi, \tag{A.13}$$

(A.4) is satisfied for all values of k. For $x = 0$, the function equals k/π, which tends to infinity. For values of $x \neq 0$, the function oscillates with increasing frequency, as $k \to \infty$, through successively positive and negative half-periods of $\sin kx$. Such infinitely rapid oscillation results in complete cancellation of all $x \neq 0$ contributions to an integral such as (A.2). Although this representation of deltafunction is not literally in accord with the model implied by (A.3), under an integral sign, its net effect is exactly the same.

The representation (A.12) can be alternatively expressed in integral form as

$$\delta(x) = \lim_{k\to\infty} \frac{1}{2\pi} \int_{-k}^{k} \mathrm{e}^{ik'x}\, \mathrm{d}k' \tag{A.14}$$

or

$$\delta(x) = \frac{1}{2\pi} \int_{-\infty}^{\infty} \mathrm{e}^{ikx}\, \mathrm{d}k. \tag{A.15}$$

This identity is utilized in Fourier analysis (cf Appendix B) and in normalization of free-particle eigenfunctions (cf. Section 5.2).

Secondly, consider the representation

$$\delta(x) = \lim_{k\to\infty} \frac{\sin^2 kx}{\pi k x^2}. \tag{A.16}$$

Again (A.4) is ensured because

$$\int_{-\infty}^{\infty} \frac{\sin^2 x}{x^2}\, \mathrm{d}x = \pi. \tag{A.17}$$

For $x = 0$, the function equals k/π, which increases without limit, while, for $x \neq 0$, the function simply approaches zero as $k \to \infty$. This representation of the deltafunction finds application in time-dependent perturbation theory (cf. Section 7.5).

The deltafunction can also be represented as the derivative of the unit step function:

$$\delta(x) = \theta'(x),$$

where

$$\theta(x) = \begin{cases} 0, & x < 0 \\ 1, & x > 0 \end{cases} \tag{A.18}$$

To demonstrate this, make use of integration by parts as follows:

$$\int_{-\infty}^{\infty} f(x)\theta'(x)\,\mathrm{d}x = [f(x)\theta(x)]_{-\infty}^{\infty} - \int_{-\infty}^{\infty} f'(x)\theta(x)\,\mathrm{d}x$$

$$= f(\infty) - \int_{0}^{\infty} f'(x)\,\mathrm{d}x = f(0) \tag{A.19}$$

showing that $\theta'(x)$ behaves as $\delta(x)$ in (A.2)†. A deltafunction will, in fact, appear whenever a discontinuous function is differentiated.

As evidenced in the variety of its possible representations, the deltafunction by itself is somewhat nebulously defined; only when it occurs under an integral sign does it acquire precise meaning. Accordingly, deltafunctions will appear in quantum theory only in intermediate stages of calculation—never as a physically significant end result.

Different quantities which produce equivalent results when they occur within an integral can be represented in terms of abstract identities. Following are several such relations involving deltafunctions.

$$\delta(-x) = \delta(x) \tag{A.20}$$

$$x\delta(x) = 0 \tag{A.21}$$

$$\delta(ax) = |a|^{-1}\delta(x) \tag{A.22}$$

$$\delta(x^2 - a^2) = (2a)^{-1}[\delta(x-a) + \delta(x+a)] \tag{A.23}$$

$$\delta[(x-a)(x-b)] = |a-b|^{-1}[\delta(x-a) + \delta(x-b)] \tag{A.24}$$

$$\int_{-\infty}^{\infty} \delta(x-a)\delta(x-b)\,\mathrm{d}x = \delta(a-b) \tag{A.25}$$

$$f(x)\delta(x-a) = f(a)\delta(x-a) \tag{A.26}$$

By means of partial integration it can be shown that

$$\int_{-\infty}^{\infty} f(x)\delta'(x)\,\mathrm{d}x = -f'(0). \tag{A.27}$$

By extension, the nth derivative the deltafunction represents a generalized function with the fundamental property

$$\int_{-\infty}^{\infty} f(x)\delta^{(n)}(x)\,\mathrm{d}x = (-1)^n f^{(n)}(0). \tag{A.28}$$

† Repeating this calculation for $\int_0^\infty$ and using (A.5) implies that $\theta(0) = \frac{1}{2}$.

Two additional identities involving $\delta'(x)$:

$$\delta'(x) = -\delta'(-x) \tag{A.29}$$

$$x\delta'(x) = -\delta(x). \tag{A.30}$$

The deltafunction can be generalized to several dimensions. In three dimensions, for example,

$$\delta(\mathbf{r} - \mathbf{r}_0) \equiv \delta(x - x_0)\delta(y - y_0)\delta(z - z_0), \tag{A.31}$$

such that

$$\int f(\mathbf{r})\delta(\mathbf{r} - \mathbf{r}_0)\, \mathrm{d}^3\mathbf{r} = f(\mathbf{r}_0). \tag{A.32}$$

For a set of n generalized coordinates $q_1 \ldots q_n$, one defines the n-dimensional deltafunction $\delta(q - q')$ having the fundamental property

$$\int f(q_1 \ldots q_n)\, \delta(q - q')\, \mathrm{d}\tau = f(q_1' \ldots q_n'). \tag{A.33}$$

At the same time,

$$\int \cdots \int f(q_1 \ldots q_n)\delta(q_1 - q_1') \ldots \delta(q_n - q_n')\, \mathrm{d}q_1 \ldots \mathrm{d}q_n = f(q_1' \ldots q_n'). \tag{A.34}$$

Since the generalized volume element is given by (cf. 3.2.5)

$$\mathrm{d}\tau = g(q_1 \ldots q_n)\, \mathrm{d}q_1 \ldots \mathrm{d}q_n, \tag{A.35}$$

the n-dimensional deltafunction has the following explicit form:

$$\delta(q - q^1) = \frac{\delta(q_1 - q_1') \ldots \delta(q_n - q_n')}{g(q_1 \ldots q_n)}. \tag{A.36}$$

In spherical polar coordinates,

$$\delta(\mathbf{r} - \mathbf{r}') = \frac{\delta(\mathbf{r} - \mathbf{r}')\delta(\theta - \theta')\delta(\phi - \phi')}{r^2 \sin^2 \theta} \tag{A.37}$$

provided that $r' \neq 0$. Alternatively, making use of the closure relation for spherical harmonics,

$$\delta(\mathbf{r} - \mathbf{r}') = \frac{\delta(r - r')}{rr'} \sum_{l=0}^{\infty} Y_{lm}^*(\theta, \phi) Y_{lm}(\theta', \phi') \tag{A.38}$$

in which r^2 has been replaced by rr' for symmetry.

The special case $\mathbf{r}' = 0$ ($r' = 0$, θ' and ϕ' arbitrary) requires some caution. Since the singular point $r = 0$ is also a boundary point of integration, the normalization condition (A.5) pertains. A factor 2 must therefore be put in on the right-hand side of (A.37):

$$\delta(\mathbf{r}) = \frac{2\delta(r)\delta(\theta - \theta')\delta(\phi - \phi')}{r^2 \sin\theta} . \tag{A.39}$$

Integrating over solid angle ($\sin\theta \, d\theta \, d\phi$), we obtain

$$\delta(\mathbf{r}) = \delta(r)/2\pi r^2 . \tag{A.40}$$

An alternative way to get this relation follows from the 3-dimensional analog of the representation (A.15):

$$\delta(\mathbf{r}) = (2\pi)^{-3} \int e^{i\mathbf{k}\cdot\mathbf{r}} \, d^3\mathbf{k} . \tag{A.41}$$

The integration is, of course, most straightforward when **k** is expressed in terms of its cartesian components. But since the result must be independent of coordinate system, one can also introduce spherical polar coordinates in **k**-space (with **r** as the polar axis), whereby

$$\delta(\mathbf{r}) = (2\pi)^{-3} \int_0^\infty \int_0^\pi \int_0^{2\pi} e^{ikn\cos\theta} k^2 \sin\theta dk d\theta d\phi . \tag{A.42}$$

Integration over angles leaves

$$\begin{aligned} \delta(\mathbf{r}) &= -\frac{i}{4\pi^2 r}\left(\int_0^\infty k\, e^{ikr}\, dk - \int_0^\infty k\, e^{-ikr}\, dk\right) \\ &= -\frac{1}{4\pi^2 r}\frac{d}{dr}\int_{-\infty}^{\infty} e^{ikr}\, dk = -\frac{1}{2\pi}\frac{\delta'(r)}{r}. \end{aligned} \tag{A.43}$$

Thus, using (A.30), we arrive again at (A.40).

One must be careful in applications not to confuse the vector deltafunction $\delta(\boldsymbol{r})$ with the radial deltafunction $\delta(r)$. Dimensionally, $\delta(\boldsymbol{r})$ has units of r^{-3}, $\delta(r)$ of r^{-1}.

The deltafunction can also be utilized in other branches of mathematical physics to unify formalism for continuous and discrete distributions. Compare, for example, the expression for the coulombic potential of a point charge (at $\mathbf{r} = \mathbf{r}_0$):

$$\Phi(\mathbf{r}) = \frac{q}{|\mathbf{r} - \mathbf{r}_0|} \tag{A.44}$$

with that of a continuous charge distribution:

$$\Phi(\mathbf{r}) = \int \frac{\rho(\mathbf{r}')}{|\mathbf{r} - \mathbf{r}'|} d^3\mathbf{r}. \tag{A.45}$$

Equation (A.44) reduces to a special case of (A.45) with

$$\rho(\mathbf{r}') = q\delta(\mathbf{r}' - \mathbf{r}_0). \tag{A.46}$$

Appendix B

Fourier Analysis

A pure harmonic function of x can be represented in the form

$$f(x) = f e^{ikx}. \tag{B.1}$$

The amplitude f can be taken as a complex quantity

$$f = |f| e^{i\alpha} \tag{B.2}$$

such that α determines the absolute phase, i.e.

$$f(x) = |f| e^{i(kx+\alpha)}. \tag{B.3}$$

In many cases, the physical significance of $f(x)$ resides only in its *real part*

$$\text{Re } f(x) = |f| \cos(kx + \alpha) \tag{B.4}$$

although the complex exponential form might be carried along for computational convenience. Under very general conditions, an arbitrary function of x in the domain $(-\infty, \infty)$ can be represented as a superposition of harmonic components by means of a *Fourier integral*†:

$$f(x) = \int_{-\infty}^{\infty} g(k) e^{ikx} dk. \tag{B.5}$$

When $g(k)$ is a discrete function of k, say,

$$g(k) = g_n \delta(k - 2\pi n/a) \quad n = 0, \pm 1, \pm 2 \ldots \tag{B.6}$$

then (B.5) reduces to a complex Fourier series

$$f(x) = \sum_{n=-\infty}^{\infty} g_n e^{i2\pi nx/a}. \tag{B.7}$$

The function $g(k)$ is called the *Fourier transform* of $f(x)$. To solve (B.5) ex-

† The Fourier integral exists in the ordinary sense if $f(x)$ is piecewise continuous and absolutely integrable ($\int_{-\infty}^{\infty} |f(x)| dx$ exists). See, for example, E. C. Titchmarsh, "Introduction to the Theory of Fourier Integrals, Oxford University Press, 1937. Within the domain of generalized functions, the integrability restriction can be relaxed (cf Lighthill, *op cit.*).

plicitly for $g(k)$, change the variable of integration to k', multiply by e^{-ikx} and integrate over x:

$$\int_{-\infty}^{\infty} f(x)\, e^{-ikx}\, dx = \int_{-\infty}^{\infty} g(k')\left[\int_{-\infty}^{\infty} e^{i(k'-k)x}\, dx\right] dk' \qquad \text{(B.8)}$$

By application of (A.15), the bracket is equivalent to $2\pi\delta(k' - k)$. We arrive thereby at the *Fourier inversion formula*

$$g(k) = \frac{1}{2\pi}\int_{-\infty}^{\infty} f(x)\, e^{-ikx}\, dx. \qquad \text{(B.9)}$$

Some simple Fourier transforms are listed in Table B.

A more symmetrical way of defining the pair of transform functions is as follows:†

$$f(x) = \frac{1}{\sqrt{(2\pi)}}\int_{-\infty}^{\infty} g(k)\, e^{ikx}\, dk$$

$$g(k) = \frac{1}{\sqrt{(2\pi)}}\int_{-\infty}^{\infty} f(x)\, e^{-ikx}\, dx. \qquad \text{(B.10)}$$

Although this formulation is preferable, there do arise instances in which the representation (B.5) is the most natural starting point. If the Fourier transform is defined according to (B.10) rather than (B.9), $g(k)$ for specific functions should be multiplied by a factor $(2\pi)^{1/2}$. These are given explicitly in the third column of Table B.

A useful and interesting result concerns the Fourier transform of a gaussian:

$$gk) = \frac{1}{\sqrt{(2\pi)}}\int_{-\infty}^{\infty} e^{-x^2/2}\, e^{-ikx}\, dx \qquad \text{(B.11)}$$

By completing the square in the exponent, we obtain

$$g(k) = e^{-k^2/2}\frac{1}{\sqrt{(2\pi)}}\int_{-\infty}^{\infty} e^{-(x+ik)^2/2}\, dx = e^{-k^2/2} \qquad \text{(B.12)}$$

Thus the gaussian function $e^{-x^2/2}$ is invariant under Fourier transformation ‡

† Another alternative which avoids prefactors of 2π entirely, is based on the conjugate variable $\not{k} \equiv k/2\pi$. Accordingly,

$$f(x) = \int_{-\infty}^{\infty} g(\not{k})\, e^{i2\pi \not{k} x}\, d\not{k}$$

$$g(\not{k}) = \int_{-\infty}^{\infty} f(x)\, e^{-i2\pi \not{k} x}\, dx.$$

‡ Another function which shares this property is $x^{1/2}J_{-1/4}(x^2/2)$ (cf. Morse and Feshbach, *op cit.* p 484).

TABLE B. Some Fourier transforms

$f(x)$	$g(k) = \frac{1}{2\pi}\int_{-\infty}^{\infty} f(x)\,e^{-ikx}\,dx$	$g(k) = \frac{1}{\sqrt{(2\pi)}}\int_{-\infty}^{\infty} f(x)\,e^{-ikx}\,dx$
$af(x)$	$ag(k)$	
$f(ax)$	$a^{-1}g(k/a)$	
$f(x+a)$	$e^{ika}g(k)$	
$f'(x)$	$-ikg(k)$	
$f''(x)$	$-k^2g(k)$	
$xf(x)$	$ig'(k)$	
$x^2f(x)$	$-g''(k)$	
$g(x)$	$(2\pi)^{-1}f(-k)$	$f(-k)$
1	$\delta(k)$	$(2\pi)^{\frac{1}{2}}\delta(k)$
e^{ik_0x}	$\delta(k-k_0)$	$(2\pi)^{\frac{1}{2}}\delta(k-k_0)$
$\delta(x-a)$	$(2\pi)^{-1}e^{-ika}$	$(2\pi)^{-\frac{1}{2}}e^{-ika}$
$e^{-x^2/2}$	$(2\pi)^{-\frac{1}{2}}e^{-k^2/2}$	$e^{-k^2/2}$
$e^{-x^2/2a^2}$	$(2\pi)^{-\frac{1}{2}}ae^{-a^2k^2/2}$	$a\,e^{-a^2k^2/2}$

More generally,

$$\frac{1}{\sqrt{(2\pi)}}\int_{-\infty}^{\infty} e^{-x^2/2a^2}\,e^{-ikx}\,dx = ae^{-a^2k^2/2} \tag{B.13}$$

so that

$$FT\,e^{-x^2/2a^2} = a\,e^{-a^2k^2/2}. \tag{B.14}$$

This applies as well when a is complex, provided only that $\mathrm{Re}\,(a^2) > 0$.

Some additional results, not difficult to prove (based on the representation B.10). If $f(x)$ is a normalized function in x-space then its transform is likewise normalized in k-space. More generally

$$\int_{-\infty}^{\infty} |f(x)|^2\,dx = \int_{-\infty}^{\infty} |g(k)|^2\,dk \tag{B.15}$$

known as Parseval's theorem. Two formulas of use in the calculation of momentum expectation values:

$$-i\int_{-\infty}^{\infty} f^*(x)\,f'(x)\,dx = \int_{-\infty}^{\infty} |g(k)|^2 k\,dk \tag{B.16}$$

and

$$-\int_{-\infty}^{\infty} f^*(x)f''(x)\,dx = \int_{-\infty}^{\infty} |g(k)|\,k^2\,dk. \tag{B.17}$$

Finally, an inequality mathematically equivalent to the Heisenberg uncertainty relation:

$$\int_{-\infty}^{\infty} |f(x)|^2(x - x_0)^2 \, dx \int_{-\infty}^{\infty} |g(k)|^2(k - k_0)^2 \, dk \geqslant \tfrac{1}{4} \tag{B.18}$$

assuming $f(x)$ (hence also $g(k)$) is normalized and where

$$x_0 \equiv \int_{-\infty}^{\infty} |f(x)|^2 x \, dx, \quad k_0 \equiv \int_{-\infty}^{\infty} |g(k)|^2 k \, dk. \tag{B.19}$$

The inequality can be condensed to

$$\Delta x \Delta k \geqslant \tfrac{1}{4} \tag{B.20}$$

in which Δx, Δk represent the root-mean-square deviations.

Fourier integrals can straightforwardly be generalized to any number of dimensions. In three dimensions, for example,

$$f(\mathbf{r}) = \int g(\mathbf{k}) \, e^{i\mathbf{k}\cdot\mathbf{r}} \, d^3\mathbf{k} \tag{B.21}$$

$$g(\mathbf{k}) = (2\pi)^{-3} \int f(\mathbf{r}) \, e^{-i\mathbf{k}\cdot\mathbf{r}} \, d^3\mathbf{r}. \tag{B.22}$$

A common pair of conjugate variables in Fourier analysis are time t and frequency ω. When $f(t)$ represents the output or signal, $g(\omega)$ gives the corresponding frequency spectrum. It is conventional to reverse the signs in the exponential factors, whereby

$$f(t) = \int_{-\infty}^{\infty} g(\omega) \, e^{-i\omega t} \, d\omega \tag{B.23}$$

$$g(\omega) = \frac{1}{2\pi} \int_{-\infty}^{\infty} f(t) \, e^{i\omega t} \, dt. \tag{B.24}$$

One reason for doing this is so that space-time Fourier integrals such as

$$f(\mathbf{r}, t) = \iint g(\mathbf{k}, \omega) \, e^{i(\mathbf{k}\cdot\mathbf{r} \quad \omega t)} \, d^3\mathbf{k} \, d\omega \tag{B.25}$$

can be expressed in covariant notation:

$$f(x) = \int g(k) \, e^{ik_\nu x_\nu} \, d^4k, \tag{B.26}$$

with $x_4 = ict$ and $k_4 = i\omega/c$. Moreover, the exponential basis functions in (B.25) coincide with the usual expression for plane waves.

INDEX